Inositol Phosphates
and Derivatives

A C S   S Y M P O S I U M   S E R I E S   **463**

Inositol Phosphates and Derivatives

Synthesis, Biochemistry, and Therapeutic Potential

Allen B. Reitz, EDITOR

R.W. Johnson Pharmaceutical Research Institute

Developed from a symposium sponsored
by the Division of Carbohydrate Chemistry
at the 200th National Meeting
of the American Chemical Society,
Washington, D.C.,
August 26–31, 1990

American Chemical Society, Washington, DC 1991

Library of Congress Cataloging-in-Publication Data

Inositol phosphates and derivatives: synthesis, biochemistry, and therapeutic potential / Allen B. Reitz, editor.

  p.   cm.—(ACS symposium series; 463).

"Developed from a symposium sponsored by the Division of Carbohydrate Chemistry at the 200th National Meeting of the American Chemical Society, Washington, D.C., August 26–31, 1990."

Includes bibliographical references and indexes.

ISBN 0–8412–2086–7: $59.95

  1. Inositol   phosphate—Congresses.   2. Inositol   phosphate—Derivatives—Congresses.

  I. Reitz, Allen Bernard, 1956– . II. American Chemical Society. Division of Carbohydrate Chemistry. III. American Chemical Society. Meeting (200th: 1990: Washington, D.C.) IV. Series.

QP772.I5I58   1991
615.7—dc20

                                                    91–17716
                                                    CIP

The paper used in this publication meets the minimum requirements of American National Standard for Information Sciences—Permanence of Paper for Printed Library Materials, ANSI Z39.48–1984. ∞

A00000462476Q

ACS Symposium Series

M. Joan Comstock, *Series Editor*

1991 ACS Books Advisory Board

Foreword

THE ACS SYMPOSIUM SERIES was founded in 1974 to provide a medium for publishing symposia quickly in book form. The format of the Series parallels that of the continuing ADVANCES IN CHEMISTRY SERIES except that, in order to save time, the papers are not typeset, but are reproduced as they are submitted by the authors in camera-ready form. Papers are reviewed under the supervision of the editors with the assistance of the Advisory Board and are selected to maintain the integrity of the symposia. Both reviews and reports of research are acceptable, because symposia may embrace both types of presentation. However, verbatim reproductions of previously published papers are not accepted.

Contents

ix

About the Cover

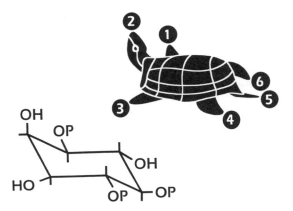

It wasn't possible to include both pieces of art on the cover, therefore they are printed here. The turtle representation of an inositol phosphate was generously provided by B. W. Agranoff and the artwork by Thomas Ford–Holevinski.

Steric relationships are conveniently visualized by employing the image of a turtle to represent the *myo*-inositol molecule. For example, in D-*myo*-inositol-1-phosphate, the head represents the axial hydroxyl at D2, while the four limbs and tail comprise the five equatorial groups. Viewed from above, the right front leg of the turtle is at D1 and proceeding counterclockwise, the head is D2, etc.

Preface

INOSITOL PHOSPHATES AND THEIR DERIVATIVES are remarkably versatile compounds, crucial for the transduction of information in living organisms. As with the formation of c-AMP by adenylate cyclase, phospholipase C-catalyzed cleavage of membrane-bound phosphatidylinositol triphosphate (PIP_2) into *myo*-inositol 1,4,5-triphosphate and diacylglycerol is a keenly important means of cellular signaling. Indeed, 28 of 74 receptors listed in the 1991 "Receptor Nomenclature Supplement" of *Trends in Pharmacological Sciences* are reported to be linked to an increase of PIP_2 turnover following receptor activation. Key aspects of research in the general area of inositol phosphates and derivatives are highlighted in the book, especially from the viewpoint of synthetic organic chemistry, with contributions from biochemistry and pharmacology.

Organic and medicinal chemists have synthesized the natural inositol phosphates to demonstrate unambiguously their structure. Additionally, these chemists have prepared a host of unnatural inositol phosphates, isosteres, and related derivatives in a search for selective and useful enzyme modulators.

If the inositol phosphates and the pathways that create them are a ubiquitous means by which cells respond to extracellular stimuli, then how can one hope to obtain therapeutically useful selectivity for one system over another? There are a number of reasons to believe that this second-messenger pathway, like others, can be favorably manipulated in vivo. Enzymes have different tissue selective isoforms, such as phospholipase C, for which five isoforms are known. Receptors and enzymes have heterogeneous distributions. Other organ- or disease-specific factors may conspire to make selectivity possible. As our knowledge of fundamental biochemistry and pharmacology grows in this area, so also will our ability to conceive of approaches to intervene with advantage.

Chemists have used a mixture of standard and novel methods to prepare inositol phosphates and derivatives. The synthetic challenges are formidable and include issues of stereocontrol, resolution of enantiomers, regioselectivity of reactions such as phosphorylations, and protecting group manipulations.

Overall, this book provides a glimpse into the chemistry required to prepare the inositol phosphates and derivatives and into the uses to which such molecules are being put as enzyme inhibitors, biochemical tools, and potential pharmacological agents. This book is intended as a resource for scientists involved in the field of inositol phosphates and derivatives in academia and industry, as well as a broad reference for organic, medicinal, and carbohydrate chemists and biochemists who are generalists with a passing interest.

Acknowledgments

I am grateful to the international group of authors whose chapters appear in this volume. I admire the diversity of the research they conducted, and I wish them the best of success in all future ventures. I hope this book will facilitate scientific interchange among these contributors and others, so that full use will be made of new discoveries related to the organic chemistry, biochemistry, and pharmacology of the inositol phosphates.

Walter W. Zajac, Jr., of Villanova University and Anthony S. Serianni of the University of Notre Dame were among the helpful colleagues from the Division of Carbohydrate Chemistry of the American Chemical Society. Financial support was also provided by the donors of the Petroleum Research Fund, which is administered by the American Chemical Society; the R.W. Johnson Pharmaceutical Research Institute, a Johnson and Johnson Company; and Merck Sharp and Dohme Research Laboratories.

I especially appreciate the assistance and suggestions of Roy Gigg and Joseph P. Vacca, who are authors of chapters in this book. The engaging turtle representation of an inositol phosphate shown on the cover was generously provided by B. W. Agranoff and the artwork by Thomas Ford–Holevinski. Even though *myo*-inositol looks like a turtle, research has advanced in this area with the speed of a hare. One would hope that the *myo*-inositol turtle is of the mutant ninja variety so we can press on in the future. Special thanks are due also to Frank Eisenberg, Jr., who briefly came out of retirement to coauthor the first chapter, and whom I wish many years of skiing and travel.

I applaud the efforts of all the reviewers who examined drafts of the enclosed chapters. I thank the R.W. Johnson Pharmaceutical Research Institute for the time and support they gave in the editing process. A. Maureen Rouhi, of the ACS Books Department, provided the needed insight and thoroughness required to bring this volume to print.

ALLEN B. REITZ
R.W. Johnson Pharmaceutical Research Institute
Spring House, PA 19477–0776

March 11, 1991

Chapter 1

Biochemistry, Stereochemistry, and Nomenclature of the Inositol Phosphates

R. Parthasarathy[1] and F. Eisenberg, Jr.[2]

[1]Department of Psychiatry, University of Tennessee Health Science Center, Memphis, TN 38105
[2]6028 Avon Drive, Bethesda, MD 20814

With the intense interest created by the inositol phospholipids a need has arisen for a review of the stereochemistry and nomenclature of their hydrolytic products, the *myo*-inositol phosphates. From elementary considerations of molecular geometry the concepts of conformation, configuration, chirality, and the *meso* compound are developed and how these concepts lead to a consistent nomenclature is explained. A new nomenclature applicable mainly to this family of compounds, including the adoption of the symbol Ins, is presented. These principles are examined in the light of enzymatic reactions of *myo*-inositol leading to chiral products and of published methods for the laboratory synthesis of chiral *myo*-inositol phosphates involved in second messenger function.

The need for a review of these topics was first recognized at about the time the role of inositol phospholipids in cellular signalling was demonstrated (*1*). With that revelation the steady but quiet activity surrounding *myo*-inositol and its congeners erupted into an avalanche of published papers. The multitude of workers newly attracted to this field were to encounter not only the subtleties of inositol stereochemistry but also confusion and even error in the codification of the rules of nomenclature. As a result many of those errors continue in the literature. We hope that this review which is based in part on the review of 1986 (*2*) and in part on the recent presentation of this subject (*3*) will help correct erroneous notions that persist and bring investigators in the field up to date.

A Source of Second Messengers

Although the *myo*-inositol molecule has engaged the interest of organic chemists for more than a century, the most exciting development has occurred in its biochemistry

0097–6156/91/0463–0001$06.00/0

with the discovery that the inositol phospholipids are the precursors of second messengers in receptor-mediated intracellular Ca^{2+} mobilization and protein phosphorylation. Events leading to that discovery, however, began thirty-odd years ago with the observation by Hokin and Hokin (4) of the increased turnover of phosphatidylinositol in exocrine pancreas stimulated with acetylcholine. The turnover of phosphatidate also increased but not that of the major membrane phospholipids, phosphatidylcholine, phosphatidylethanolamine, and sphingomyelin. Thereafter the observation was repeated in a vast array of systems exposed to a variety of agonists, but for the next twenty years its role in cell physiology remained an enigma.

Turnover of the Inositol Phospholipids

The chemistry of the turnover of the inositol phospholipids is summarized in Figure 1. Although the synthetic phase of the cycle had been worked out early on (Figure 1c and Figure 2)(5,6), the breakdown of phosphatidylinositol remained ambiguous with three possibilities considered: reversal of synthesis (Figure 2); phospholipase D type of hydrolysis with the formation of phosphatidate and free myo-inositol (not shown); and phospholipase C type giving diacylglycerol and myo-inositol 1-phosphate (Figure 1d) which would then be hydrolyzed to free myo-inositol and inorganic phosphate (Figure 1b). The key to this puzzle was the discovery that the lithium ion was a specific inhibitor of myo-inositol 1-phosphatase (7). In the presence of Li^+, stimulated breakdown of phosphatidylinositol led to the accumulation of myo-inositol 1-phosphate, clearly pointing to the phospholipase C mechanism, followed by hydrolysis of myo-inositol 1-phosphate in the absence of Li^+ (8). In the last decade a role for this turnover has emerged with the discovery, foreshadowed by earlier work (9,10), that it is the agonist-sensitive breakdown of phosphatidyl-inositol 4,5-bisphosphate that is the crucial step in the turnover giving rise transiently to $Ins(1,4,5)P_3$, the second messenger for Ca^{2+} mobilization within the cell. Concurrent with the breakdown of all the phosphoinositides (11), diacylglycerol is liberated and plays its second messenger role as the transient activator of protein kinase C (12). Apart from rapid appearance at low concentration, the other criterion which defines a signalling role is the rapid disappearance of the compound; mechanisms for the removal and reutilization of both second messengers are also shown in Figure 1. Owing to the vast output of experimental work confirming these observations these seminal discoveries seem to be universal.

Enantiomers of myo-Inositol 1-Phosphate

Figure 1 comprises two regions marked de novo and turnover. In the de novo pathway (13) labeled (a), its brevity belying its complexity, glucose 6-phosphate is isomerized to $Ins(3)P$ which is then hydrolyzed by a specific phosphatase to free myo-inositol and inorganic phosphate (Figure 3); this reaction along with dietary sources provides a net pool of myo-inositol. In the turnover cycles free myo-inositol is generated by the same enzyme from $Ins(1)P$. These monophosphates are different forms of myo-inositol 1-phosphate, differing only optically as indicated by the rotational signs (-) and (+), respectively. The unusual circumstance of finding

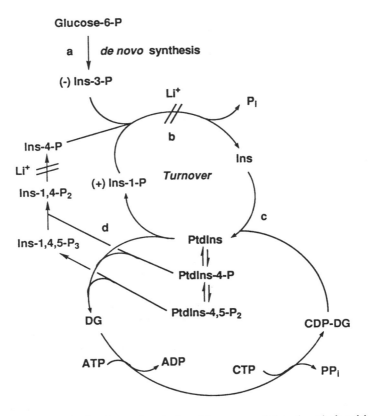

Glucose-6-P

a │ *de novo* synthesis

(-) Ins-3-P

Figure 1. Biosynthesis of *myo*-inositol and turnover of the phosphoinositides. Ptd, phosphatidyl; Ins, *myo*-inositol numbered counterclockwise; P, phosphate; CTP, cytidine triphosphate; DG, diacyl glycerol.

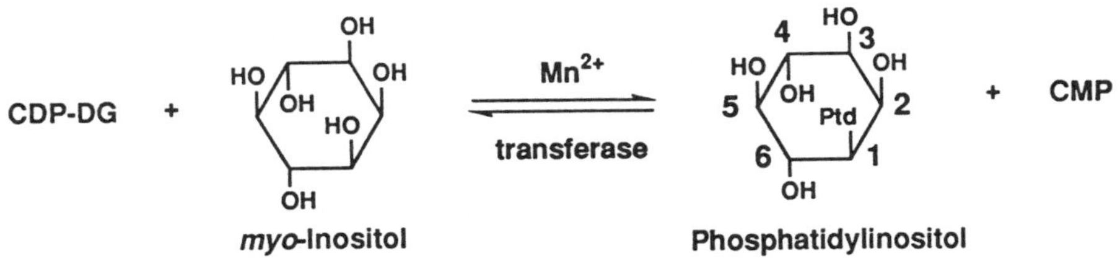

CDP-DG + *myo*-Inositol $\xrightarrow[\text{transferase}]{Mn^{2+}}$ Phosphatidylinositol + CMP

Figure 2. Reversible biosynthesis of phosphatidylinositol. (Reproduced with permission from ref. 2. Copyright 1986 *The Biochemical Journal*.)

optical isomers, or enantiomers, of the same compound in a metabolic sequence invites an inquiry into the stereochemistry of both free and esterified *myo*-inositol and the influence it has had on the evolution of the rules of nomenclature in this class of compounds. In a practical application of stereochemical principles, Sherman and his colleagues *(14)* assessed the relative contribution of the two pathways to the *myo*-inositol pool in bovine brain. Since both enantiomers are equal substrates for the specific phosphatase, their relative contributions can not be ascertained by measurement of free *myo*-inositol. With the addition of Li$^+$, the phosphatase is inhibited and now by chromatographic resolution of the accumulated enantiomers these investigators found a preponderance of Ins(1)P, the enantiomer arising from the breakdown of phosphatidylinositol. Phospholipase C activity is thus the source of brain *myo*-inositol. Hoffmann-Ostenhof *et al. (15)* similarly assessed the relative contributions of competing reactions in a study of stereospecific enzymatic methylation of *myo*-inositol. The application of stereochemical principles is clearly vital to biochemical studies.

Stereochemistry and Molecular Geometry

It must be stated at the outset that stereochemistry, despite its long history, is an ever developing discipline, aptly characterized as *"in statu nascendi" (16)*. Even among experts there are sharp theoretical differences *(17)* and no set of rules of nomenclature has satisfied the versatility required of it. Our interest is confined to a relatively small class of compounds embodying the inositol structure whose stereochemistry can be understood in simple terms and with a few rules. For more advanced treatment of the subject the reader is directed to the work of Mislow and Siegel *(18)*, among others.

To understand stereochemistry it is necessary to examine molecular geometry, a fact first recognized by Pasteur who separated by hand enantiomorphic forms of tartrate salts by their opposite crystal habits. The development of stereochemistry followed from the tetravalency of carbon and the assumption that the four valences were uniformly disposed in space. A regular tetrahedron could then represent the carbon atom which if bound to four different ligands was said to be asymmetric or stereogenic. The same structure can exist in the form of its non-superimposable mirror image or enantiomer. Because of its application to organic chemistry, stereochemistry tends to be associated with the carbon atom only; in addition, the phosphorus atom is of stereochemical importance as attested by current studies of the mechanism of the reactions of phosphoinositide-specific phospholipase C *(19)*.

The Six-Carbon Ring. Since *myo*-inositol consists of a ring of six carbon atoms joined by single bonds, we begin with a simplified perspective diagram of a ring of six regular tetrahedra (Figure 4). To accommodate the spatial requirements peculiar to this structure we arrange the solids in their most favorable, *i.e.*, least energetic, strain-free state known as the chair form, in which alternating members lie above and below the plane common to their bases. A line connecting the centers of each solid through an apex reveals the familiar puckered ring of *myo*-inositol (Figure 5). Emanating perpendicularly from the centers (Figure 4) are the six axial valences alternating above and below the ring; lines joining the centers to the remaining

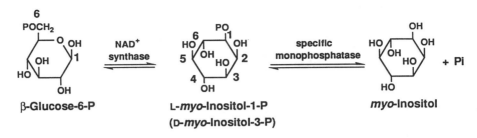

Figure 3. *De novo* biosynthesis of *myo*-inositol.

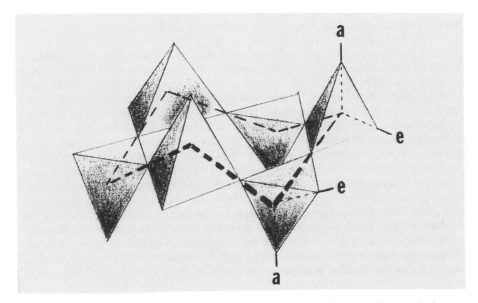

Figure 4. Chair form. a, axial; e, equatorial. (Reproduced with permission from ref. 2. Copyright 1986 *The Biochemical Journal*.)

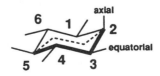

Figure 5. *Myo*-inositol, conformational diagram, with hydroxyls omitted. (Adapted from ref. 2.)

apexes and alternating up and down at about 20° represent the six equatorial valences. By rotating each tetrahedron another equally probable strain-free form results in which axial and equatorial valences are interchanged. Although it is necessary to strain the ring, this change can be effected without rupture of the connections between the solids. Such a chair-chair interconversion is known as a conformational change and the two forms are called conformational isomers or conformers. Binding of a hydrogen and a hydroxyl to each carbon tetrahedron produces the hexahydroxycyclohexanes or cyclohexanehexols, known more familiarly as the inositols. Although containing six OH groups, the inositols are not hexitols as they are sometimes erroneously referred to; as isomers of $C_6H_{12}O_6$ the inositols are at the same level of oxidation as the hexoses. Hexitols, $C_6H_{14}O_6$, are reduced hexoses; the inositols have no such analogs.

Stereoisomers

These six secondary hydroxyl groups can be arranged among nine stereoisomeric forms with axial groups varying from none to three (Figure 6). Referred to by their prefixes they are *scyllo-* (no axial OH); *myo-* (1); *neo-, epi-*, D-*chiro-*, and L-*chiro-* (2); *cis-, muco-*, and *allo-* (3). Those with three axial OH groups can undergo chair-chair interconversions readily; those with less than three are less readily interconverted. *allo-*Inositol is unique in that its conformer is also its enantiomer. An optically inactive mixture (50/50) of enantiomers is known as a racemate. *allo-*Inositol is such a racemate. Although several of the inositols are found in nature, far and away the most abundant is *myo-*inositol and it is the only stereoisomer to occur in the phospholipids. Why nature has selected the *myo* structure to build on is conjectural; the most cogent argument is its uniqueness in containing a single axial OH group which may impose just the right number of stereoselective sites for enzyme attack. *scyllo-*Inositol with no axial OH group is devoid of stereoselective sites. Notwithstanding its small size and compact carbocyclic structure, the *myo-*inositol molecule serves as a paradigm for the illustration of conformation, chirality, configuration, and optical and geometric isomerism.

Numbering of the *myo*-Inositol Ring

The heavy lines in Figure 5 indicate that C3 is nearest the viewer with the axial OH group at C2 above the plane of the ring. To ensure that the ring is numbered consistently on the side bearing the axial OH group, Agranoff (20) has contrived a whimsical mnemonic in the form of a turtle with the head erect at C2, the splayed legs representing C1, C3, C4, C6 and the tail C5. By convention the numbering of *myo-*inositol is clockwise viewed from above the ring.

Haworth Projection

A projection originally applied to carbohydrates by Haworth to highlight configurational relationships is shown in Figure 7 for *myo-*inositol. It represents a more compact form of perspective drawing than Figure 5, with the puckered ring projected onto a plane and the hydroxyl groups placed vertically above and below the plane according to their up or down direction in Figure 5. The OH groups at C1, C2, C3, C5 are *cis* to each other as are those at C4, C6. The two groups in turn are *trans* to each other. The C3, C4 edge is closest to the viewer and numbering is clockwise viewed from above the ring. *myo*-Inositol is unique in having a single cluster of three vicinal (adjacent) *cis* OH groups; the middle one is axial at C2 and the transannular *cis* OH group, equatorial at C5.

From Figures 5 and 7 it is apparent that the *myo*-inositol molecule can be divided into mirror image halves as shown by the trace of the plane of symmetry between C2 and C5. In this connection there is a curious anomaly in the Haworth projection of *allo*-inositol, the result of the failure of Haworth projections to distinguish axial from equatorial bonds. In Figure 6 *allo*-inositol is shown to consist of a mixture of enantiomers but Figure 8 is contradictory in showing a spurious plane of symmetry in one or the other enantiomer according to the Haworth projection when in fact no plane exists as seen in the chair form.

Vicitrans Rules

This discrepancy points up the difficulty in inferring three-dimensional conformations from Haworth projections. Although deriving the Haworth diagram from the chair form is obvious, the reverse is not. To facilitate this translation we have devised what we call vicitrans (*vicinal, cis, trans*) rules: Rule 1, in a vicinal *cis* pair of OH groups, if one is axial the other must be equatorial, and conversely; Rule 2, in a vicinal *trans* pair both must be axial or equatorial. The rules can be tested with Figure 7; starting clockwise from C1 vicinal OH pairs follow the order *cis, cis, trans, trans, trans, trans*. If the OH group at C1 is arbitrarily called equatorial, then by Rule 1 the OH group at C2 is axial and continuing clockwise, by Rule 2 the remaining OH groups are equatorial. This order e, a, e, e, e, e is unique to *myo*-inositol. If C2 had been called axial instead the order would have been a, e, a, a, a, a, the less probable conformer of *myo*-inositol. D-*chiro*- and L-*chiro*-inositols can not be distinguished from each other by these rules.

The *Meso* Compound

myo-İnositol (Figure 7) consists of a ring of six hydroxymethylene groups four of which, C1, C3, C4, C6, are stereogenic. Although chemically identical those at C1 and C3 are geometrically opposite; the same applies to C4 and C6. Since each group compensates geometrically for its paired opposite, the net stereogenicity of these four groups is zero. And since the groups at C2 and C5 lie in the plane of symmetry, they contribute no chirality to the molecule. *myo*-Inositol, then, is an achiral molecule which is divisible along the line connecting C2 and C5 into compensating chiral halves. Compounds of this kind are called *meso* compounds. Contrary to what is often seen in the literature *myo*-inositol is neither D nor L; configurational prefixes are not applicable to *meso* compounds.

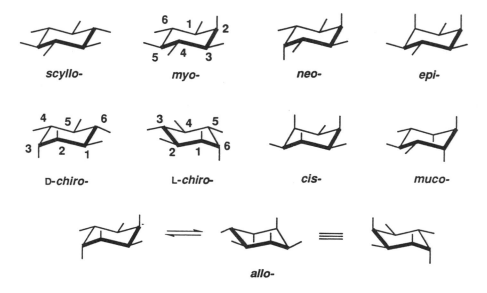

Figure 6.   The inositol stereoisomers with hydroxyls omitted.

Figure 7.   *Myo*-inositol, Haworth projection.

Figure 8.   *Allo*-inositol.

Chirality of *myo*-Inositol Derivatives

Substitution of a phosphate (or any other group) on an OH at a stereogenic position will disturb the symmetry of *myo*-inositol and render the derivative chiral; substitution at C2 or C5 will not disturb symmetry and the derivative will remain achiral and *meso*. *myo*-Inositol 1-phosphate is chiral but both *myo*-inositol 2-phosphate and *myo*-inositol hexakisphosphate (phytic acid) are *meso* compounds. They also are often given configurational prefixes where none should be applied.

Figure 9 illustrates the enantiomers of *myo*-inositol 1-phosphate and shows that they are in fact non-superimposable mirror images. The D enantiomer [called Ins(1)P in Figure 1] is derived from the breakdown of phosphatidylinositol and the L enantiomer [Ins(3)P in Figure 1] from the isomerization of glucose 6-phosphate. Their numbering is in accordance with the convention that newly substituted positions be assigned to the lowest possible locant (numbered position)(*21*). To meet that requirement the two compounds must be numbered in opposite directions, clockwise for L and counterclockwise for D, on the side of the ring bearing the axial OH group. Before 1968 the assigned configurations were the reverse of Figure 9 and readers referring to the literature of that era should be aware of the change. Originally, configuration of *myo*-inositol derivatives was assigned according to the rules of carbohydrate nomenclature in which the highest numbered stereogenic carbon specified configuration, but in 1968 the rule was changed to the lowest numbered stereogenic carbon, C1 of *myo*-inositol. For the L enantiomer rotating the molecule in the plane of the ring and viewing the C6-C1 edge with C6 at the top, the phosphate will point left, hence L configuration. A simple hand maneuver will make this clear. Extend the right hand palm up and thumb bent; rotate the hand 90°. Calling the index finger C6 and the thumb (phosphate) C1 the direction of the phosphate is left.

Configurational Relationships

Historically the *myo*-inositol 1-phosphates are important in the comprehension of the stereochemistry and nomenclature of the *myo*-inositol phosphates in general. In their classical proof of structure of phosphatidylinositol (Figure 2) Ballou and Pizer (*22*) hydrolyzed soybean phosphoinositides and isolated an enantiomer of *myo*-inositol 1-phosphate of unknown configuration. To establish the configuration of this compound they also prepared the opposite enantiomer from galactinol (Figure 10), a compound of known L configuration. Accordingly, the phospholipid-derived enantiomer and the parent phosphatidylinositol were assigned D configuration. By optical comparison with these enantiomers L configuration was assigned to the *myo*-inositol 1-phosphate produced in the enzymatic cyclization of glucose 6-phosphate (Figure 3)(*13*). Since all work cited in this paragraph was done before 1968, the published configurational notation is opposite to that reported here.

Recognition of Chirality

The enantiomers of *myo*-inositol 1-phosphate reflect the enantiotopic regions (thermodynamically identical but geometrically compensating loci) within the parent *myo*-inositol.   Although obviously not separable these regions are nevertheless resolvable by stereospecific enzymatic attack giving rise to enantiomeric substitution products.   A non-stereoselective attack on a *meso* compound can not effect this resolution; acid or alkali treatment of *myo*-inositol 2-phosphate, causing migration of the phosphate group, leads to a racemic mixture of enantiomers of *myo*-inositol 1-phosphate.   Only stereoselective attack can recognize enantiotopic regions in a *meso* compound.   It is to be expected then that enzymatic reactions with *myo*-inositol will result in optically pure products.

Stereospecificity

The general principle of stereoselective attack on a *meso* compound will be illustrated in the examples which follow.   It is interesting and probably related to nature's predilection for the *myo* isomer that the C1-C6 edge of the molecule is the preferred site of attack.

Phosphorylation.   L-*myo*-inositol 1-phosphate (Figure 9) is the sole product of phosphorylation of *myo*-inositol by ATP catalyzed by a plant kinase (*23*).   This is the only known direct route for the phosphorylation of *myo*-inositol; the indirect routes are shown in Figure 1.

Galactosylation.   Two reactions are known for the galactosylation of *myo*-inositol (Figure 10).   UDPgalactosyl transferase attacks *myo*-inositol at C1 to give galactinol (*24*), L-1-O-galactosylinositol; beta-galactosidase attacks at C6 but to conform to the lowest-locant rule the product, beta-galactinol (*25*), is named D-4-O-galacto-sylinositol.   It is interesting that each enzyme attacks the same side of the ring but at sites that differ diastereotopically (stereoisomeric but not mirror images), a reflection of a chemical difference between C1 and C6.   In contrast to enantiotopic sites that require a stereoselective agent for recognition, diastereotopic sites can be differentiated by any chemical agent or physical method.

Dephosphorylation.   Phytic acid, *myo*-inositol hexakisphosphate, is hydrolyzed by a 1-phytase from *Pseudomonas* and a 6-phytase from wheat bran (*26*).   Like the galactosylating enzymes the phytases are enantioselective but differ from each other in selecting diastereotopic positions.   By the lowest-locant rule the product of 1-phytase hydrolysis is D-*myo*-inositol 1,2,4,5,6-pentakisphosphate and of a 6-phytase, L-*myo*-inositol 1,2,3,4,5-pentakisphosphate.   These compounds are diastereomers.

Oxidation.   An important enzyme found in all plants and mammalian kidney is *myo*-inositol oxygenase which catalyzes the atmospheric oxidation of *myo*-inositol to D-glucuronic acid via *myo*-1-inosose as shown in Figure 11 (*27*).   Not only is the C1-C6 edge the preferred site of addition and removal of substituents but of ring opening as well.   Glucuronic acid is a precursor of ascorbic acid and plant polysaccharides.

Biosynthesis.   The C1-C6 edge is also the site of ring closure as seen in the irreversible isomerization of beta-D-glucose 6-phosphate to L-*myo*-inositol 1-

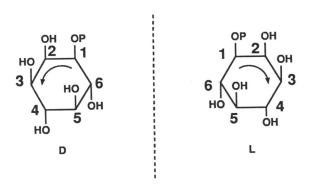

Figure 9. Enantiomeric *myo*-inositol 1-phosphates.

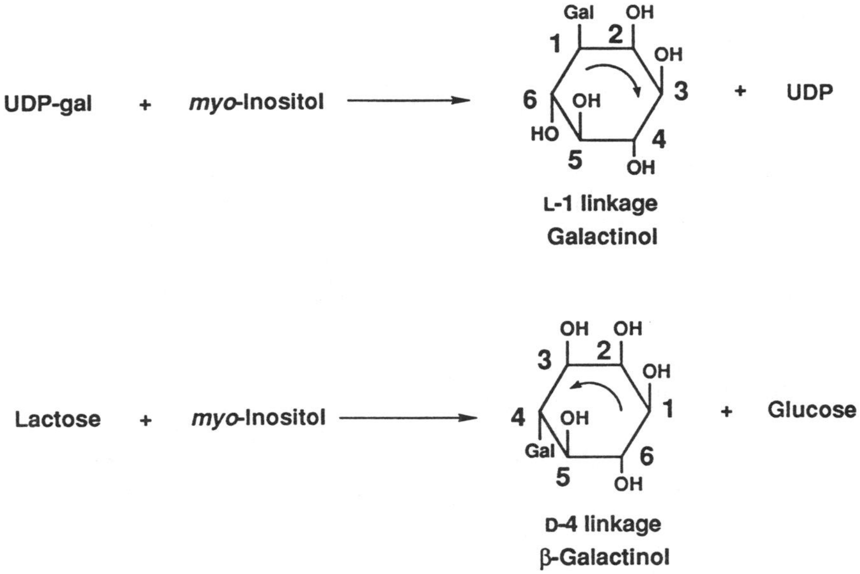

UDP-gal + *myo*-Inositol ⟶ 6 ... 3 + UDP

L-1 linkage
Galactinol

Lactose + *myo*-Inositol ⟶ 4 ... 1 + Glucose

D-4 linkage
β-Galactinol

Figure 10. Galactosylation of *myo*-inositol. Gal, galactosyl.

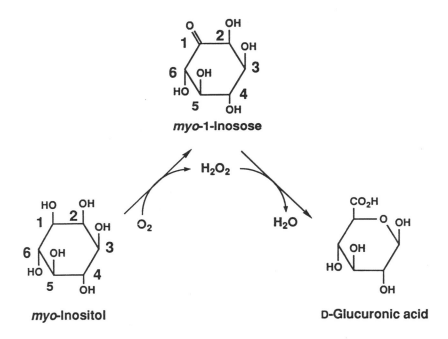

Figure 11. Mechanism of action of *myo*-inositol oxygenase. (Reproduced with permission from ref. 2. Copyright 1986 *The Biochemical Journal*.)

phosphate (Figure 3) (*13*). The first step in this reaction is an intramolecular aldol condensation but the synthase does not fit either the Class I (Schiff base mechanism) or Class II (metal ion catalysis) categories of aldolases (*28*) and probably operates by a base-catalyzed reaction mechanism (*29*). The substrate glucose 6-phosphate is activated for C1-C6 ring closure through oxidation by tightly bound NAD^+ to a short-lived species which then cyclizes to relatively long-lived but enzyme-bound *myo*-2-inosose 1-phosphate (*30,31*). This intermediate is then reduced to L-*myo*-inositol 1-phosphate by bound NADH generated in the first step. Although the short-lived species can be formally represented as 5-ketoglucose 6-phosphate there is no evidence for the finite existence of that compound. There are many sites of stereospecific selectivity in this reaction sequence eventuating unerringly in the *myo* structure. These involve the transfer of H at C4 of the coenzyme, the retention of H at C6 of glucose 6-phosphate and the selection of C-O bonds of trigonal (carbonyl) carbon at all stages of the reaction (*2*). These studies have all been performed with mammalian synthases; yeast synthase which catalyzes the same overall reaction has been analyzed from the biochemical, molecular, and genetic standpoints with the ultimate goal the dissociation of the protein into regions which can catalyze separately the partial reactions described above (*32*).

The second step is hydrolysis by a phosphatase different from either acid or alkaline phosphatase and specific for either enantiomer of *myo*-inositol 1-phosphate, behavior that departs from the optical specificity usually displayed by enzymes. That the enzyme selects equatorial phosphates is borne out by its full activity with L-*chiro*-inositol 3-phosphate (*13*) and Ins(4)*P* (Figure 1) and its inactivity toward the axial phosphates in *myo*-inositol 2-phosphate and alpha-glucose 1-phosphate (*13*). *Myo*-inositol 1-phosphatase serves as an analytical reagent in the measurement of *myo*-inositol biosynthesis (*33*). The enzyme is inhibited by Li^+ (Figure 1).

Phosphatidylation. The notable exception to the rule stated earlier that the C1-C6 edge is the preferred site of enzymatic attack is the phosphatidylation of *myo*-inositol at C3 in the reversible synthesis of phosphatidylinositol (Figure 2).   From the lowest-locant rule phosphatidylinositol is named systematically phosphatidyl-1-D-*myo*-inositol, the numbering of the ring counterclockwise.

The Phosphoinositides

Phosphatidylinositol, the major membrane phosphoinositide, is accompanied by its phosphorylated derivatives, phosphatidylinositol 4-phosphate and phosphatidylinositol 4,5-bisphosphate. As shown in Figure 1 these give rise on hydrolysis by phospho-inositide-specific phospholipase C to Ins(1)*P*, Ins(1,4)P_2, and Ins(1,4,5)P_3, respectively, the last now established as a second messenger in agonist-sensitive Ca^{2+} mobilization.   Diacylglycerol, the remainder of the molecule, is the second messenger for protein phosphorylation by protein kinase C.   A series of newly discovered phosphoinositides has appeared whose functions remain unclear; these are phosphatidylinositol 3-phosphate (*34*), phosphatidylinositol 3,4-bisphosphate (*35,36*), and phosphatidylinositol 3,4,5-trisphosphate (*37,38*).   Whether this last is a precursor of the other participant in Ca^{2+} mobilization Ins(1,3,4,5)P_4, is unclear but it has been shown that this tetrakisphosphate is the product of phosphorylation of

Ins(1,4,5)P_3 (*39*). The structures of three of these compounds of current interest are shown in Figure 12.

Other Functions of Phosphatidylinositol

Phosphatidylinositol also appears in glycosylated forms to provide anchorage for hydrophilic proteins in cell membranes and by virtue of their susceptibility to hydrolysis by phosphoinositide-specific phospholipase C to provide for quick release of these proteins from the membranes (*40*). The linkage at one end of the branched sugar moiety is to C6 of *myo*-inositol of the phospholipid and the other end of the oligosaccharide is linked through mannose 6-phosphate esterified to the C-terminal alpha carboxyl of protein-bound aspartic acid. These inositol glycolipids also serve as precursors of insulin mediators (*41*), which in general have been shown to contain linked galactose, glucosamine, and *myo*-inositol. A more recently discovered mediator contains instead mannose, galactosamine, and D-*chiro*-inositol (see Figure 6 for structure) (*42*). This finding suggests that *myo*-inositol is not the only stereoisomer found in inositol phospholipids. The chirality of the *chiro*-inositol was established as D by comparison with authentic D- and L-*chiro*-inositols by GC/MS analysis on a capillary column of Chirasil Val III, a chiral resin originally designed for the resolution of enantiomeric amino acids. This is the same method that was used in the identification of Ins(1)P in bovine brain mentioned earlier (*14*).

myo-Inositol Phosphates

Flowing from the phosphoinositides is an ever widening stream of hydrolytic products and their phosphorylated progeny (*1*). Of sixty-six possible isomers of phosphorylated *myo*-inositol (*43*) sixteen have been found in all known pathways. It seems likely that with the exploration of other plant and animal systems the list may well expand. Most of these are products of agonist-sensitive phosphoinositide hydrolysis and of these most are of D configuration as expected from their anteced-ents, a few are of L configuration, and some are achiral.

Nomenclature of the *myo*-Inositol Phosphates. Because of this large number of known *myo*-inositol phosphates with probably more to come and to avoid the confusion arising out of the vagaries of configuration dictated by the lowest-locant rule, it was felt by people in the field that these compounds should constitute a family with its own rules of nomenclature. Accordingly, the Nomenclature Committee of the International Union of Biochemistry promulgated a set of proposals and recommendations to meet these needs (*21*). They propose the adoption of the stereospecific numbering (*sn*) system used with glycerol derivatives and as a corollary the relaxation of the lowest-locant rule. Implicit in this proposal is the numbering of all *myo*-inositol phosphates counterclockwise since that is the direction of numbering in the phosphoinositides. Only those few presently known compounds of L configuration would be affected by this change; L-*myo*-inositol 1-phosphate would become *myo*-inositol 3-phosphate; L-*myo*-inositol 1,6-bisphosphate, *myo*-inositol 3,4-bisphosphate; and L-*myo*-inositol 1,4,5,6-tetrakisphosphate, *myo*-inositol 3,4,5,6-tetrakisphosphate. Although the D prefix is used in the published rules the *sn* system obviously does not require it. The second recommendation is the adoption of the symbol Ins to represent *myo*-inositol numbered counterclockwise. *myo*-Inositol 1-phosphate becomes Ins(1)*P* and its enantiomer becomes Ins(3)*P*. The drawback in these rules is that they obscure not only enantiomeric relationships but the absence of chirality as well. The symbol Ins must imply only counterclockwise numbering and not necessarily D configuration; otherwise, there would be a conflict in its application to *meso* compounds like Ins(1,3)*P*$_2$ and other achiral phosphates. With those caveats in mind the symbol Ins should serve the phosphoinositide community well.

Synthetic Methods for *myo*-Inositol Phosphates and Related Compounds

We conclude with a summary of synthetic methods for *myo*-inositol phosphates and analogs developed to provide adequate amounts of material for further investigation of this expanding field. Some of these methods will be described in detail in the chapters which follow. They all illustrate the principles of stereochemistry set out in this review and in general they follow two courses: (a) racemate formation by non-stereoselective attack on a *meso* compound followed by resolution to yield pure enantiomers, and (b) preparation of pure enantiomers from chiral precursors without the need for resolution.

(a). *myo*-Inositol is the usual *meso* starting compound. Hydroxyl groups are protected with an array of ethers, acetals, and esters and then differentially deprotected to expose those groups destined for phosphorylation, for which a large variety of newly developed phosphorylating agents is available. Either before or after phosphorylation the resulting racemate is resolved by various chiral agents, including an enzymatic resolution. The diastereomers are purified by crystallization or chromatography and the chiral agent and protective groups removed, liberating the pure chiral phosphate. In this way Ins(4)*P* *(44)* and its enantiomer *(44)*, Ins(1,4)*P*$_2$ *(44)* and enantiomer *(44)*, Ins(1,3,4)*P*$_3$ *(45)*, Ins(1,4,5)*P*$_3$ *(44-49)* and enantiomer *(44,46,47)*, and Ins(1,3,4,5)*P*$_4$ *(45,50)* have been prepared.

Instead of *myo*-inositol as precursor benzene can also be converted into the inositol structure. By oxidation with *Pseudomonas putida* benzene gives rise to *cis*-1,2-dihydroxycyclohexa-3,5-diene, a *meso* compound *(51)*. The substrate is achiral with no geometrically opposite pairs to differentiate but by introduction of oxygen at one of the double bonds compensating stereogenic hydroxymethylene groups are generated. These are then blocked and the product is oxidized further to give another pair of OH groups; the product is now a racemic mixture as expected from a non-stereoselective attack on a *meso* compound. These enantiomers are resolved, the remaining double bond oxidized to give six hydroxyl groups and with suitable methylation and deprotection D- and L-pinitol are produced. The pinitols are methyl ethers of D- and L-*chiro*-inositol, respectively, and serve as precursors of other inositol stereoisomers.

(b). In these procedures chiral starting materials are used and no racemization takes place at any stage of synthesis.

As described in a subsequent chapter L-quinic acid (L-1,3,4,5-tetrahydroxycyclohexane 1-carboxylic acid), a widely distributed plant product, can be converted in several steps into the *myo* structure and ultimately to the *myo*-inositol phosphates and their analogs.

Chiral *myo*-inositol phosphates have been prepared in high yield from D-pinitol (D-5-O-methyl-*chiro*-inositol) and L-quebrachitol (L-2-O-methyl-*chiro*-inositol), available in abundance from plant sources. Both of the *chiro*-inositols are numbered clockwise from a *cis* pair of OH groups above the ring (Figure 13). The letter a̲ indicates a single pair of vicinal axial OH groups. See Figure 6 for the chair form of the *chiro*-inositols to visualize the axial OH groups.

For the synthesis of Ins(1)P, L-quebrachitol was first subjected to oxidoreductive inversion at the C1 axial OH group to give the *myo*-inositol structure in which C1 is derived from C5 of L-quebrachitol. By demethylation followed by suitable protection of OH groups the OH at C1 was phosphorylated (*52*).

In a highly sophisticated synthesis (*53*) of Ins(1,4,5)P_3 and Ins(3,5,6)P_3 (the D and L enantiomers of *myo*-inositol 1,4,5-trisphosphate, respectively) D-pinitol and L-quebrachitol (Figure 13) were first demethylated to D- and L-*chiro*-inositol, respectively. The free *chiro*-inositol was benzylated at the vicinal *cis* OH groups (L5, L6; D3, D4) and three of the remaining OH groups were benzoylated leaving one axial OH free, which was inverted by triflate displacement giving the *myo* structure. Which of the two axial OH groups is inverted is inconsequential since the *chiro*-inositols with their unique pair of vicinal axial OH groups can be folded on the line a̲ to give superimposable halves. These are not mirror image halves as in *myo*-inositol but thermodynamically and geometrically identical halves; the fold line does not locate a plane of symmetry as these molecules can not be divided symmetrically. Removal of the benzyl groups gave L-*myo*-inositol 1,2,4-tribenzoate from D-pinitol and D-*myo*-inositol 2,3,6-tribenzoate from L-quebrachitol. Phosphorylation and saponification gave Ins(1,4,5)P_3 from D-pinitol and Ins(3,5,6)P_3 from L-quebrachitol.

It is interesting that L-quebrachitol is the precursor of the configurationally opposite Ins(1)P and Ins(3,5,6)P_3 (D and L, respectively) in the two methods just described. Common to both procedures is the change from the *chiro* to the *myo* structure requiring inversion of one of the axial OH groups. Then by manipulation of the blocking agents the desired position for phosphorylation is reached. In the course of synthesis the chirality of the precursor *chiro*-inositol was conserved in the intermediate *myo*-inositol derivative, leading finally to a chiral phosphorylated *myo*-inositol. By observing the lowest-locant rule L-quebrachitol yielded D-*myo*-inositol 1-phosphate (*52*) (counterclockwise numbering) and L-*myo*-inositol 1,4,5-trisphosphate (*53*) (clockwise numbering); by relaxation of the rule and using the symbol Ins the same products are Ins(1)P and Ins(3,5,6)P_3, respectively. Since the *chiro*-inositols are of considerable interest in the synthesis of analogs of the *myo*-inositol phosphates (*54*) and appeared in earlier work with insulin mediators (*42*), the heuristic value of a detailed stereochemical examination of these procedures can not be overestimated.

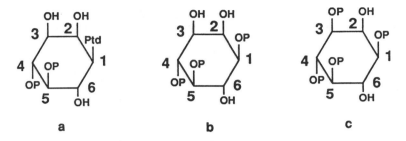

Figure 12.   Structures of *myo*-inositol derivatives important in agonist-sensitive Ca^{2+} mobilization.   (a) PtdIns(4,5)P_2; (b) Ins(1,4,5)P_3; (c) Ins(1,3,4,5)P_4.

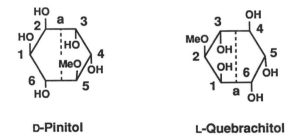

D-Pinitol   **L-Quebrachitol**

Figure 13.   Precursors of synthetic chiral *myo*-inositol phosphates.

Literature Cited

1. Berridge, M. J.; Irvine, R. F. *Nature* **1989**, *341*, 197-205.
2. Parthasarathy, R.; Eisenberg, F., Jr. *Biochem. J.* **1986**, *235*, 313-322.
3. Eisenberg, F., Jr. *Abstract No. 48*, Div. Carbohydrate Chem., 200th ACS National Meeting, Washington, D.C., Aug. 26-31, 1990.
4. Hokin, M. R.; Hokin, L.E. *J. Biol. Chem.* **1953**, *203*, 967-977.
5. Agranoff, B. W.; Bradley, R. M.; Brady, R. O. *J. Biol. Chem.* **1958**, *233*, 1077-1083.
6. Paulus, H.; Kennedy, E. P. *J. Biol. Chem.* **1960**, *235*, 1303-1311.
7. Hallcher, L. M.; Sherman, W. R. *J. Biol. Chem.* **1980**, *255*, 10896-10901.
8. Hokin-Neaverson, M.; Sadeghian, K. *J. Biol. Chem.* **1984**, *259*, 4346-4352.
9. Durell, J.; Garland, J. T. In *Cyclitols and Phosphoinositides: Chemistry, Metabolism, and Function*; Eisenberg, F., Jr., Ed.; *Ann. N. Y. Acad. Sci.* **1968**, *165*, 743-754.
10. Abdel-Latif, A. A.; Akhtar, R. A.; Smith, J. P. In *Cyclitols and Phosphoinositides*; Wells, W. W.; Eisenberg, F., Jr. Eds.; Academic Press: New York, NY, 1978; pp 121-143.
11. Majerus, P. W.; Connolly, T. M.; Deckmyn, H.; Ross, T. S.; Bross, T. E.; Ishii, H.; Bansal, V.; Wilson, D. B. *Science* **1986**, *234*, 1519-1526.
12. Nishizuka, Y. *Nature* **1988**, *334*, 661-665.
13. Eisenberg, F., Jr. *J. Biol. Chem.* **1967**, *242*, 1375-1382.
14. Sherman, W. R.; Leavitt, A. L.; Honchar, M. P.; Hallcher, L. M.; Phillips, B. E. *J. Neurochem.* **1981**, *36*, 1947-1951.
15. Hoffmann-Ostenhof, O.; Pittner, F.; Koller, F. In *Cyclitols and Phosphoinositides*; Wells, W. W.; Eisenberg, F., Jr., Eds.; Academic Press: New York, NY, 1978; pp 233-248.
16. Dodziuk, H.; Mirowicz, M. *Tetrahedron: Asymmetry* **1990**, *1*, 171-186.
17. O'Loane, J. K. *Chem. Rev.* **1980**, *80*, 41-61.
18. Mislow, K.; Siegel, J. *J. Am. Chem. Soc.* **1984**, *106*, 3319-3328.
19. Lin, G.; Burnett, C. F.; Tsai, M.-D. *Biochemistry* **1990**, *29*, 2747-2757.
20. Agranoff, B. W. *Trends Biochem. Sci.* **1978**, *3*, N283-N285.
21. IUB Nomenclature Committee *Biochem. J.* **1989**, *258*, 1-2.
22. Ballou, C. E.; Pizer, L. I. *J. Am. Chem. Soc.* **1960**, *82*, 3333-3335.
23. Loewus, M. W.; Sasaki, K.; Leavitt, A. L.; Munsell, L.; Sherman, W. R.; Loewus, F. A. *Plant Physiol.* **1982**, *70*, 1661-1663.
24. Frydman, R. B.; Neufeld, E. F. *Biochem. Biophys. Res. Commun.* **1963**, *12*, 121-125.
25. Kuo, C.-H.; Wells, W. W. *J. Biol. Chem.* **1978**, *253*, 3550-3556.
26. Irving, G. C. J. In *Inositol Phosphates: Their Chemistry, Biochemistry, and Physiology*, Cosgrove, D. J., Ed.; Elsevier, Amsterdam, 1980; pp 85-98.
27. Naber, N. I.; Hamilton, G. A. *Biochim. Biophys. Acta* **1987**, *911*, 365-368.
28. Maeda, T.; Eisenberg, F., Jr. *J. Biol. Chem.* **1980**, *255*, 8458-8464.
29. Sherman, W. R.; Hipps, P. P.; Mauck, L. A.; Rasheed, A. In *Cyclitols and Phosphoinositides*; Wells, W. W.; Eisenberg, F., Jr. Eds.; Academic Press, New York, NY, 1978; pp 279-295.

30. Eisenberg, F., Jr.; Maeda, T. In *Inositol and Phosphoinositides: Metabolism and Regulation*; Bleasdale, J. E.; Eichberg, J.; Hauser, G., Eds.; Humana Press, Clifton, NJ, 1985; pp 3-11.

31. Eisenberg, F., Jr.; Parthasarathy, R. In *Methods in Enzymology*; Conn, P. M.; Means, A. R., Eds.; Academic Press, New York, NY, 1987, Vol. 141; pp 127-143.

32. Dean-Johnson, M.; Henry, S. A. *J. Biol. Chem.* **1989**, *264*, 1274-1283.

33. Eisenberg, F., Jr.; Parthasarathy, R. In *Methods of Enzymatic Analysis*; Bergmeyer, H. U., Ed.; Verlag Chemie, Weinheim, 1984, Third Edition, Vol. VI; pp 371-375.

34. Stephens, L.; Hawkins, P.T.; Downes, C. P. *Biochem. J.* **1989**, *259*, 267-276.

35. Traynor-Kaplan, A. E.; Thompson, B. L.; Harris, A. L.; Taylor, P.; Omann, G. M.; Sklar, L. A. *J. Biol. Chem.* **1989**, *264*, 15668-15673.

36. Yamamoto, K.; Lapetina, E. G. *Biochem. Biophys. Res. Commun.* **1990**, *168*, 466-472.

37. Traynor-Kaplan, A. E.; Harris, A. L.; Thompson, B. L.; Taylor, P.; Sklar, L. A. *Nature* **1988**, *334*, 353-356.

38. Vadnal, R. E.; Parthasarathy, R. *Biochem. Biophys. Res. Commun.* **1989**, *163*, 995-1001.

39. Lee, S. Y.; Sim, S. S.; Kim, J. W.; Moon, K. H.; Kim, J. H.; Rhee, S. G. *J. Biol. Chem.* **1990**, *265*, 9434-9440.

40. Doering, T. L.; Masterson, W. J.; Hart, G. W.; Englund, P. T. *J. Biol. Chem.* **1990**, *265*, 611-614.

41. Romero, G.; Luttrell, L.; Rogol, A.; Zeller, K.; Hewlett, E.; Larner, J. *Science* **1988**, *240*, 509-511.

42. Larner, J.; Huang, L. C.; Schwartz, C. F. W.; Oswald, A. S.; Shen, T.-Y.; Kinter, M.; Tang, G.; Zeller, K. *Biochem. Biophys. Res. Commun.* **1988**, *151*, 1416-1426.

43. Majerus, P. W.; Connolly, T. M.; Bansal, V. S.; Inhorn, R. C.; Ross, T. S.; Lips, D. L. *J. Biol. Chem.* **1988**, *263*, 3051-3054.

44. Vacca, J. P.; deSolms, S. J.; Huff, J. R.; Billington, D. C.; Baker, R.; Kulagowski, J. J.; Mawer, I. M. *Tetrahedron* **1989**, *45*, 5679-5702.

45. Yu, K.-L.; Fraser-Reid, B. *Tetrahedron Lett.* **1988**, *29*, 979-982.

46. Vacca, J. P.; deSolms, S. J.; Huff, J. R. *J. Am. Chem. Soc.* **1987**, *109*, 3478-3479.

47. Stepanov, A. E.; Runova, O. B.; Schlewer, G.; Spiess, B.; Shvets, V. I. *Tetrahedron Lett.* **1989**, *30*, 5125-5128.

48. Ozaki, S.; Watanabe, Y.; Ogasawara, T.; Yoshihisa, K.; Shiotani, N.; Nishii, H.; Matsuki, T. *Tetrahedron Lett.* **1986**, *27*, 3157-3160.

49. Liu, Y.-C.; Chen, C.-S. *Tetrahedron Lett.* **1989**, *30*, 1617-1620.

50. Watanabe, Y.; Oka, A.; Shimizu, Y.; Ozaki, S. *Tetrahedron Lett.* **1990**, *31*, 2613-2616.

51. Ley, S. V.; Sternfeld, F. *Tetrahedron* **1989**, *45*, 3463-3476.

52. Akiyama, T.; Takechi, N.; Ozaki, S. *Tetrahedron Lett.* **1990**, *31*, 1433-1434.

53. Tegge, W.; Ballou, C. E. *Proc. Natl. Acad. Sci. U.S.A.* **1989**, *86*, 94-98.

54. Carless, H. A. J.; Busia, K. *Tetrahedron Lett.* **1990**, *31*, 1617-1620.

RECEIVED March 18, 1991

Chapter 2

Phosphoinositides and Their Stimulated Breakdown

B. W. Agranoff[1] and S. K. Fisher[2]

[1]Department of Biological Chemistry,[2] Department of Pharmacology, University of Michigan, Ann Arbor, MI 48109–0720

Myo-inositol containing lipids play a central role in signal transduction whereby ligands on the cell surface initiate specific physiological responses. Biochemical and pharmacological aspects of the measurement of stimulated phosphoinositide turnover are reviewed, together with other recent developments concerning the function of inositol-containing lipids.

The cyclitols, especially *myo*-inositol, and the *myo*-inositol containing lipids have enjoyed renewed interest as a result of the demonstration that they are components of a major cellular signal transduction system (*1,2,3*). Binding of any of a number of ligands to specific receptors on the outer leaflet of the eukaryotic cell plasma membrane results in the increased breakdown of inositol-containing phospholipids (phosphoinositides) in the inner plasma membrane leaflet, which then leads to an elevation of the intracellular calcium ion concentration $[Ca^{2+}]_i$, and an increase in phosphorylation of proteins. These in turn mediate depolarization, contraction, secretion, mitosis or differentiation, the nature of the response depending on the nature of the cell type and of the receptor that is activated. The stimulated breakdown of inositol lipids thus constitutes the basis of a major second messenger system whereby chemical signals outside the cell are transduced across the cell membrane.

Interest in the regulation of $[Ca^{2+}]_i$ extends beyond its second messenger functions. There exist pathological conditions, such as brain hypoxia, that can lead to an unregulated elevation of $[Ca^{2+}]_i$ which in turn may threaten cell viability as a result of depletion of ATP, the activation of Ca^{2+}-mediated proteases and lipases, or other deleterious events. In such instances, it may be desirable to block elevation of $[Ca^{2+}]_i$ in order to maintain cell viability (*4*).

For these various physiological and unphysiological considerations, it is of interest to develop agents by which one may alter

0097–6156/91/0463–0020$06.00/0

the intracellular phosphoinositide-mediated second messenger system pharmacologically, in the hope that therapeutic agents may emerge. One should note *a priori*, however, that the utility of manipulating intracellular second messenger systems may be questionable; there are by now a wide variety of ligands, including neurotransmitters and hormones, that exert their known biological effects via the receptor-activated phosphoinositide second messenger system (Table I). In this respect, the brain and other neural-related tissues are particularly enriched in receptors that operate through this pathway. Given the broad spectrum of ligands that share the system, it would appear unlikely that one could produce a pharmacological effect with the desired selectivity by its modification. It can be noted by way of analogy that while there is an enormous literature on the regulation of the cyclic AMP-mediated signal transduction pathway, as well as a number of available agents that will inhibit 3'-5' cyclic AMP formation or metabolism, one has yet to successfully manipulate it pharmacologically for the treatment of a specific pathological condition *in vivo*. There may nevertheless be cause for optimism regarding the possible development of a useful interventive agent in the case of the phosphoinositide-mediated signal transduction pathway on the basis of the apparent mechanism of action of Li^+ in the treatment of manic depressive disorders (5). A favored hypothesis is based on the demonstration that Li^+ blocks intracellular inositol phosphate breakdown, as is discussed below. One may hope that even better agents for the treatment of this mental disorder and other mental dysfunctions can be developed. This reasoning, as well as the possible regulation of neuronal $[Ca^{2+}]_i$, may account for some of the present interest and activity in the pharmaceutical industry in identifying agents that modify the phosphoinositide signal transduction pathway. While it is too early to speculate that successful new drugs will emerge, at the very least the search may provide useful experimental agents that will increase our understanding of the phosphoinositide-mediated transduction process.

The Phosphoinositides and Their Biosynthesis

In phosphatidylinositol (PI), 1,2-diacyl-*sn*-glycerol (DAG) is phosphodiesterified to the D1 position of *myo*-inositol. PI (Figure 1A) is by far the most commonly occurring inositol lipid in nature. It constitutes about 5% of the membrane phospholipid content of eukaryotic cells. The higher phosphoinositides, phosphatidylinositol 4-phosphate [PI(4)P] and phosphatidylinositol 4,5-bisphosphate [PI(4,5)P$_2$] are less abundant, together representing less than 1% of total cellular phospholipids, and are thought to be enriched in plasma membranes. PI(4)P and PI(4,5)P$_2$ have not generally been considered components of plant lipids. However, since they are key intermediates in the phosphoinositide-mediated signal transduction pathway, and that there has been recent evidence for the operation of this pathway in plants (7), we must acknowledge the significance of their presence, if not the actual amounts. Less common inositol-containing lipids, including an inositol-containing sphingolipid, will not be considered in this review.

Table I. Pharmacological Profile of Receptor-Stimulated Phosphoinositide Turnover

Adrenergic (α_{1A} and α_{1B}). *Brain, liver, smooth muscle*

Angiotensin. *Liver, adrenal cortex, anterior pituitary*

Antigen. *B-lymphocytes, T-cells, basophil leukemic cells*

Bombesin. *Pancreas, brain, fibroblasts*

Bradykinin (B_2). *Kidney tubules, brain, pituitary neuroblastoma-glioma*

Cerulein. *Pancreas*

Cholecystokinin. *Neuroblastoma*

Cholinergic (M_1, M_3, M_5, muscarinic). *Brain, parotid, pancreas, smooth muscle, gastric mucosa, neuroblastoma, astrocytoma, retina, cochlea, adrenal medulla*

Concanavalin A. *Thymocytes*

Endothelin. *Brain, smooth muscle, gliomas*

f-Met-Leu-Phe. *Neutrophils, HL-60 cells*

Glutamate (Quisqualate-metabotropic). *Brain, retina*

Histaminergic (H_1). *Brain, astrocytoma*

Nerve growth factor. *Pheochromocytoma*

Neurokinin. *Brain, anterior pituitary, retina, superior colliculus*

Neurotensin. *Neuroblastoma-glioma*

Neuropeptide Y. *Dorsal root ganglion*

Phytohemagglutinin. *T-cell leukemic cells*

Platelet activating factor. *Platelets, liver, brain*

Platelet-derived growth factor. *Fibroblasts*

Prostaglandin E_2. *Adrenal medulla*

Purinergic (P_2). *Brain, neuroblastoma, adrenal medulla*

Serotonergic ($5HT_2$ and $5HT_{1C}$). *Brain, platelets, choroid plexus, insect salivary gland, smooth muscle*

Substance P. *Parotid, brain, retina, pituitary*

Thrombin. *Platelets, fibroblasts*

Vasopressin (V_1). *Liver, brain, superior cervical ganglion*

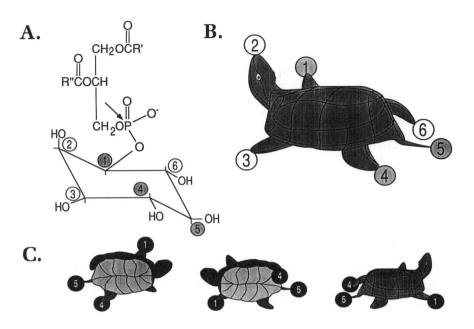

Figure 1

The structure of the phosphoinositides. **A.** Phosphatidylinositol consists of 1,2-diacyl-*sn*-glycerol phosphodiesterified to *myo*-inositol at the cyclitol D1 position. Two further phosphorylated phosphoinositides are phosphatidylinositol 4-phosphate (PI4P) and phosphatidylinositol 4,5-bisphosphate [PI(4,5)P$_2$; PIP$_2$]. The arrow indicates the site of cleavage of PIP$_2$ following receptor-ligand binding, to form diacylglycerol and inositol 1,4,5-trisphosphate [I(1,4,5)P$_3$]. **B.** Steric relationships are conveniently visualized by employing the image of a turtle to represent the *myo*-inositol molecule, in which the head represents the axial hydroxyl at C2, while the 4 limbs and tail comprise the 5 equatorial groups. Viewed from above, the right front leg of the turtle is at D1 and proceeding counterclockwise, the head is D2, etc. (ref. 6). **C.** I(1,4,5)P$_3$ "turtles." The numbered balls represent phosphomonoester groups (see also chapter by Parthasarathy and Eisenberg, this volume). Note that the turtle model facilitates identification of the molecule in different orientations. In phosphoinositides the right front leg is always phosphodiesterified to DAG.

As with the quantitatively major tissue phospholipids phosphatidylcholine (PC) and phosphatidylethanolamine (PE), PI biosynthesis begins with reduction and acylation of glycolytic intermediates, leading to the formation of phosphatidate (PA; 1,2 diacyl-*sn*-glycerophosphate; Figure 2A). Unlike PC and PE, PI synthesis does not utilize the Kennedy pathway (DAG + CDP-base); rather, PA is pyrophosphorylated by CTP to produce the liponucleotide cytidine diphosphodiacylglycerol (CDP-DAG), which then reacts with inositol to yield PI and CMP. The PI synthase reaction is highly selective for *myo*-inositol. This specificity must be dealt with in considering the possibility of introducing a cyclitol analog that would lead to cellular biosynthesis of a homologous PI analog. The putative bogus cyclitol must first be taken up by a cellular carrier system that recognizes it as authentic *myo*-inositol. Block of the synthase has not been studied extensively, although galactinol is a known inhibitor (8). PI is then phosphorylated via ATP and a PI 4-kinase, which has been termed PI kinase II (9). $PI(4)P$ is in turn phosphorylated via PIP 5-kinase to form $PI(4,5)P_2$. Specific phosphatases can break the $PI(4,5)P_2$ down to $PI(4)P$ and thence to PI. $PI(5)P$ has not been isolated from cells, an indication that PIP 5-kinase will not phosphorylate PI, and further, that PIP_2 phosphatase is selective for the 5' position and will not dephosphorylate $PI(4,5)P_2$ at the 4' position. In stimulated breakdown of $PI(4,5)P_2$ (Figure 2B), it is cleaved at the phosphodiester position by a specific Ca^{2+} activated phospholipase C, found in both membranes and cytosol. A number of such phospholipases have been purified and sequenced, and of note is the observation that despite their common function, they do not all share sequence homology (10). A recently described aminosteroid inhibitor of phospholipase C, U-73122, offers promise in studies of the enzyme's regulation (11). It is possible that in addition to those phospholipases regulated by G-proteins following receptor activation, there are degradative enzymes unrelated to the stimulated turnover of phosphoinositides. At elevated $[Ca^{2+}]_i$ concentrations, phosphoinositide-specific phospholipases C will also break down $PI(4)P$ and PI. The product of phosphoinositide cleavage in each case is DAG and, depending on whether the parent lipid is $PI(4,5)P_2$, $PI(4)P$, or PI, either $I(1,4,5)P_3$, $I(1,4)P_2$ or $I(1)P_1$, respectively. These various inositol phosphates are also possible metabolic breakdown products of $I(1,4,5)P_3$.

Recently, a PI 3-kinase, also termed PI kinase I, has been described and linked to cell transformation by oncogenes (9). The 3-kinase seems relatively unspecific, and could act on PI, $PI(4)P$ or $PI(4,5)P_2$, to produce $PI(3)P$, $PI(3,4)P_2$ or $PI(3,4,5)P_3$, respectively. Alternatively, $PI(3)P$ could be processed by PI kinase II to $PI(3,4)P_2$ and thence via the PIP 5-kinase to the postulated $PI(3,4,5)P_3$. There is evidence that this pathway exists in human platelets (12). Inositol lipids phosphorylated at the 3 position are as yet not firmly implicated in receptor-linked stimulated phosphoinositide turnover, although they have been reported to appear in non-transformed leukocytes stimulated with chemotactic peptide (13).

A. Biosynthesis

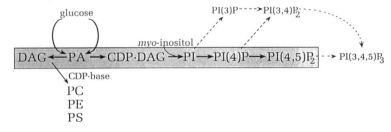

B. Stimulated Turnover

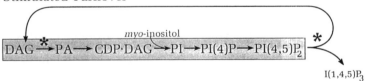

Figure 2

Comparison of *de novo* biosynthesis of phosphoinositides (A) and the regenerative cycle seen upon stimulation of receptor-linked PI-phospholipase C, as discussed in the text. The pathway for stimulated turnover (B) involves 2 additional steps indicated by the asterisks (*): DAG kinase and phosphoinositide-specific phospholipase C.

Fatty Acid Components of the Phosphoinositides

PI, PI(4)P and PI(4,5)P$_2$ share a distinctive fatty acid composition. While it is a general rule that glycerophospholipids have saturated fatty acid esterified to the 1 position and unsaturated fatty acids on the 2 position of *sn*-glycerol, there is generally a distribution of different saturated and unsaturated fatty acids at each locus. There is a much higher degree of stringency in the three inositol lipids: they are highly enriched in a single DAG species that contains only stearic acid in the 1 position and arachidonic acid (AA) in the 2 position, such that stearoyl arachidonoyl DAG usually constitutes over 80% of the total DAG species in PI, PI(4)P or PI(4,5)P$_2$. Since AA is also a precursor of a vast family of physiologically active derived lipids, the prostanoids, there has been much speculation regarding the possibility that the AA in phosphoinositides serves as a special source for prostanoid synthesis. While the phosphoinositides are the most highly enriched of the phospholipids in regard to AA content, they nevertheless do not represent a major reservoir of total cellular AA, since they are quantitatively minor phospholipids. The AA release may be an indirect result of stimulated phosphoinositide turnover, since increased $[Ca^{2+}]_i$ resulting from PI(4,5)P$_2$ breakdown, as described below, activates a phospholipase A$_2$ that releases the AA from PC (*14*).

Stimulated Breakdown of Phosphoinositides

When a ligand interacts with a phosphoinositide-linked receptor, such as when carbachol is added to intact parotid gland cells, a G-protein is activated, which leads to stimulation of a phosphoinositide-specific phospholipase C and the subsequent phosphodiesteratic cleavage of PI(4,5)P$_2$ into DAG and I(1,4,5)P$_3$ (Figure 2B). Each of the two moieties of PI(4,5)P$_2$ has second messenger actions. DAG activates protein kinase C (PKC), which in turn stimulates the phosphorylation of a number of intracellular proteins (*15*). The PKC regulatory unit binds the DAG at a site identical to that of the phorbol ester binding site, long known to have tumor-promoting actions. A variety of agents that stimulate PKC activity, including natural substances such as fatty acids and synthetic DAGs have been identified. Exogenous PKC blockers include H7 and staurosporine. Endogenous blockers of PKC include a number of sphingosine derivatives (*16*).

I(1,4,5)P$_3$ binds to a specific intracellular receptor either in the endoplasmic reticulum (ER) or plasma membrane, leading to the release of Ca^{2+} into the cytosol. The IP$_3$ receptor is particularly enriched in the cerebellum of the brain. Calmedin, a protein that has been isolated from this source and which is restricted to neural tissues, mediates the Ca^{2+} regulation of IP$_3$ binding to its receptor (*17*). A number of blockers of the IP$_3$ receptor have been identified, the most potent of which is heparin. Release of Ca^{2+} from the ER plasma membrane is also blocked by this agent.

Since I(1,4,5)P$_3$ is a chemical signaling substance, the termination of its action is as important as is its initiation. A specific inositol polyphosphate 5' phosphatase converts I(1,4,5)P$_3$ to I(1,4)P$_2$, which is inactive in stimulating Ca^{2+} release. The 5'-phosphatase is blocked by 2,3 bisphosphoglycerate. $I(1,4,5)P_3$ can alternatively be phosphorylated by a 3-kinase (distinct from the aforementioned PI 3-kinase) to form $I(1,3,4,5)P_4$. The tetrakisphosphate binds relatively poorly to the IP_3 binding site, but may in itself have action in the uptake of extracellular Ca^{2+}, and a potentiating effect on the action of $I(1,4,5)P_3$ (*1*). $I(1,3,4,5)P_4$ is quickly dephosphorylated by the 5'-phosphatase to yield $I(1,3,4)P_3$. This IP_3 isomer, unlike $I(1,4,5)P_3$, does not simulate release of $[Ca^{2+}]_i$. A non-hydrolyzable analog of $I(1,4,5)P_3$ that activates the IP_3-mediated release of Ca^{2+}, inositol 1,4,5-triphosphorothioate, has been synthesized (*18*). Its potential use as a drug is limited by its highly charged nature, which blocks its entry into normal cells. Whether the ionic charge of this agent or similar ones can be masked, for example by esterification, is a challenge for the future. Both 5-deoxy-*myo*-inositol and 5-deoxy-5-fluoro-*myo*-inositol have been synthesized with the idea that each might be incorporated into a bogus phosphoinositide (*19*). Although neither is readily taken up into cells, there is evidence that radiolabeled 5-deoxy-5-fluoro-*myo*-inositol is incorporated into PI, and to a lesser extent PIP (but not into PIP_2). This experimental approach may permit a more direct evaluation of the contribution that each of the inositol lipids plays in generation of intracellular second messengers.

Both intracellular and extracellular roles for inositol polyphosphates have been proposed. Inositol hexaphosphate (phytate) was discovered as a component of cereals over 100 years ago, and before the disease rickets became treatable with vitamin D, it was imperative to remove this substance from foods because of the rachitogenic effect of its presence in the diet, as a result of the chelation of dietary calcium in the gut. Inositol polyphosphates other than phytate, particularly those having vicinal phosphates are also effective Ca^{2+} chelating agents. $I(1,2,6)P_3$, a component of phytate hydrolysates, has found clinical application as a chelator and anti-inflammatory agent (*20*). While it has been postulated that circulating inositol polyphosphates may have biological actions (*21*), and indeed specific binding sites for IP_6 have been identified (*22*), one must first exclude the possibility that they are acting as chelators.

Cyclic Inositol Phosphates

When inositol lipids are cleaved enzymatically, varying amounts of 1:2 cyclic inositol phosphates are formed together with the non-cyclic forms. Thus, PI, PI(4)P and $PI(4,5)P_2$ will yield, on cleavage, varying amounts of $I(1:2$ cyclic$)P_1$, $I(1:2$ cyclic 4$)P_2$ and $I(1:2$ cyclic 4,5$)P_3$, respectively. The significance of the cyclic phosphates is speculative. The cyclic trisphosphate has been reported to bind to the IP_3 binding site and to release Ca^{2+}, yet is somewhat resistant to the 5'-phosphatase. It has been proposed that it thus could serve as a "long-acting" Ca^{2+}-releasing agent, but its efficacy has been called into question (*23*). Enzymatic cleavage of the cyclic ester appears to occur only at the $I(1:2$ cyclic$)P_1$ level, with the formation of the D1 phosphate (and no D2 phosphate). In designing inositol phosphate derivatives that might get into cells, the cyclic phosphate has the potential value of removing one negative charge, although the remaining charge is more highly acidic than the monophosphates, and must also be masked, if the substance is to penetrate cells. The low pK_a for the remaining charge of the cyclic phosphate is surmised from behavior of cyclic inositol phosphates on high voltage electrophoresis (B.W. Agranoff, unpublished).

Cyclization between a vicinal axial and equatorial hydroxyl can be catalyzed by heat or dehydrating agents, such as dicyclo-hexyldicarbodiimide (24). Thus, chemical cyclization of $I(1)P_1$ yields only the 1:2 cyclic ester. Cyclization of $I(2)P_1$ will yield a mixture of 1:2 and 2:3 cyclic IP_1's. Vicinal equatorial phosphates may condense to form a 7-membered cyclic pyrophosphate ring.

Resynthesis of Inositol Lipids and the Action of Lithium

A major value of Li^+ for the experimentalist is that in combination with ^3H-inositol preincubation, one can measure operation of the phosphoinositide-mediated second messenger pathway by measuring the accumulation of inositol monophosphates. This innovation, uti-lized by Berridge (4), has supplanted the earlier technique, intro-duced by the Hokins (25), which employed ^{32}P and measured the stimulated incorporation in PA and PI (Table II). Li^+ blocks the breakdown of inositol phosphates by inhibition of inositol mono-phosphate phosphatase.

I(1,4,5)P_3 released in the cleavage of PI(4,5)P_2 is eventually degraded by phosphatases to the monophosphate level, and finally to free inositol. In the presence of 1-10 mM Li^+, however, inositol monophosphate accumulates in proportion to the stimulated break-down of PI(4,5)P_2, and free inositol is not formed. The lower intra-cellular level of free inositol is then proposed to slow the resynthe-sis of PI from CDP-DAG. If indeed Li^+ blocks the synthase step by limiting the availability of inositol, then one would anticipate that CDP-DAG accumulates in Li^+-treated, ligand-stimulated cells. This has indeed been found to be the case (26,27). One could then hypo-thetically use CDP-DAG accumulation to measure the effect of Li^+, in various tissues. In contrast to measuring labeled inositol phos-phate accumulation, measurement of CDP-DAG accumulation has the potential added advantage that it indicates that phosphoinositide turnover has indeed been compromised by the presence of Li^+.

The Therapeutic Action of Li^+

That Li^+ may benefit patients with manic depressive psychosis was discovered serendipitously (28). In an experiment to test the pos-sible effects of uric acid administration, it was found that the Li^+ salt was convenient because of its relatively high solubility. Control experiments with Li salts revealed that the cation mediated the observed calming effect. Allison, Sherman and colleagues made the important observation that rats treated chronically with Li^+ had high levels of inositol monophosphates in the brain, but not in other tis-sues (29) They eventually discovered the block by Li^+ of inositol phosphate breakdown. Berridge (4) has pointed out that since the

Table II. Techniques for Measuring Activity of the Phosphoinositide-Linked Second Messenger System

Substrate	Parameters Measured
[^{32}P]Inorganic phosphate	Labeled inositol lipids and phosphatidate
[^{3}H]Inositol (+Li$^+$)	Labeled inositol phosphate
[^{3}H] or [^{14}C]Cytidine (+Li$^+$)	Labeled CDP-DAG
[^{3}H]I(1,4,5)P$_3$ or [^{32}P]I(1,3,4,5)P$_4$	Mass of I(1,4,5)P$_3$ or I(1,3,4,5)P$_4$

Indirect Approaches

[Ca^{2+}]$_i$ Elevation, DAG production, etc.

inhibition by Li^+ of the phosphatase is uncompetitive, the amount of inhibition by Li^+ will be proportional to the amount of substrate inositol phosphate available (i.e., the effect of Li^+ becomes more pronounced as inositol monophosphate concentrations increase) and this may constitute the basis of the therapeutic effect of Li^+. If manic depressive psychosis involves overstimulated neurons (admittedly a rather crude approximation of psychiatric disease), then an agent that selectively blocks signal transduction in the overstimulated cells and that has a lesser (or no) effect on normal cells could prove of therapeutic value by essentially acting as a "calcistat." There is an immediate objection to the proposal that Li^+ retards resynthesis of PI by limiting free inositol: inositol is present in brain cells in high (millimolar) concentrations. It is therefore difficult to imagine that intracellular inositol would be reduced to a sufficiently low level to block PI resynthesis. A possible counter argument in favor of the idea that Li^+ could induce an intracellular inositol deficiency is the possibility that the PI synthase enzyme in plasma membranes differs from that in the ER by having a relatively high K_m for inositol.

Li^+ is by no means an ideal drug; it is not invariably effective, and is accompanied by undesirable side effects, including tremor. We may therefore anticipate considerable sustained interest in this area because of the possibility that an agent superior to Li^+ may be found.

PI-Linked Proteins

A wide variety of proteins in a number of organisms are tethered to the outer membrane leaflet of cells via PI (30,31,32). In each instance, PI is linked via a glycoside bridge at the D6 position of myo-inositol to non-acetylated glucosamine, thence via 3 mannosyl residues to the hydroxyl group of ethanolamine by a phosphodiester linkage. The amino group of ethanolamine is amide-linked to the terminal carboxyl group of one of a large variety of proteins which is thus anchored to the membrane surface. In the case of the variant surface glycoprotein (VSG) of trypanosomes, linkage is to a terminal aspartate, and there is also a tetragalactosyl side chain linked to a mannose residue. Rat brain acetylcholinesterase is linked to the anchor ethanolamine via a terminal glycine, and there is an additional phosphoethanolamine linked to mannose. In the case of the PI anchors, the stearoyl arachidonoyl substituents are no longer seen. In VSG, dimyristoyl DAG is found, while in acetylcholinesterase, octadecanol is ether-linked to the 1 position of sn-glycerol, and 22:6 is found in the sn-2 position. It has been proposed, as for PI, that the phosphatidyl moiety of the anchor DAGs initially contains prevalent (16:0, 18:0, 18:1, etc.) fatty acids and that extensive remodeling then occurs via deacylation-reacylation reactions. Of particular interest are reports that D-chiro-inositol rather than myo-inositol may in some instances be present in PI anchors. It has been suggested that in diabetes, a phosphatidylinositol glycan may in fact be an insulin second messenger (33). The observation by Larner that D-chiro-inositol is normally present in the urine and is decreased in diabetes (34), lends further independent evidence that D-chiroinositol anchored proteins occur (*32*). The biosynthesis of D-*chiro*-inositol, and the effects of the regulation of PI anchor cleavage remain interesting issues for the future.

Acknowledgments

This work was supported by NIH grants NS 23831 (SKF), NS 15413 (BWA), and MH 42652 (BWA and SKF).

Literature Cited

1. Berridge, M. J.; Irvine, R. F. *Nature* **1989**, *341*, 197-205.
2. Fisher, S. K.; Agranoff, B. W. *J. Neurochem.* **1987**, *48*, 999-1017.
3. Rana, R. S.; Hokin, L. E. *Physiol. Rev.* **1990**, 70, 115-164.
4. Siesjö, B. K. *Ann. N. Y. Acad. Sci.* **1988**, *522*, 638-661.
5. Berridge, M. J.; Downes, C. P.; Hanley, M. R. *Cell* **1989**, *59*, 411-419.
6. Agranoff, B.W. *Trends Biochem. Sci.* **1978**, *3*, N283-N285.
7. Irvine, R. F. *Nature* **1990**, *346*, 700-701.
8. Benjamins, J.; Agranoff, B.W. *J. Neurochem.* **1969**, *16*, 513-527.
9. Whitman, M.; Downes, C. P.; Keeler, M.; Keller, T.; Cantley, L. *Nature* **1988**, *332*, 644-646.
10. Rhee, S. G.; Suh, P.-G.; Ryu, S.-H.; Lee, S. Y. *Science* **1989**, *244*, 546-550.
11. Smith, R. J.; Sam, L. M.; Justen, J. M.; Bundy, G. L.; Bala, G. A.; Bleasdale, J. E. *J. Pharmacol. Exp. Ther.* **1990**, *253*, 688-697.
12. Cunningham, T. W.; Lips, D. L.; Bansal, V. S.; Caldwell, R. K.; Mitchell, C. A.; Majerus, P. W. *J. Biol. Chem.* **1990**, *265*, 21676-21683.
13. Traynor-Kaplan, A. E.; Harris, A. L.; Thompson, B. L.; Taylor, P.; Sklar, L. A. *Nature* **1988**, *334*, 353-356.
14. Lapetina, E. G. In *Phosphoinositides and Receptor Mechanisms*, J. W. Putney, Jr., Ed.; Alan R. Liss, Inc., New York, 1986, Vol. 7; pp 271-286.
15. Nishizuka, Y. *Nature* **1989**, 334, 661-665.
16. Hannun, Y. A.; Loomis, C. R.; Merrill, A. H.; Bell, R. M. *J. Biol. Chem.* **1986**, *261*, 12604-12609.
17. Danoff, S. K.; Supattapone, S.; Snyder, S. H. *Biochem. J.* **1988**, 254, 701-705.
18. Nahorski, S. R.; Potter, B. V. L. *Trends Pharmacol. Sci.* **1989**, *10*, 139-144.
19. Moyer, J. D.; Reizes, O.; Ahir, S.; Jiang, C.; Malinowski, N.; Baker, D. C. *Mol. Pharmacol.* **1988**, *33*, 683-689.
20. Claxson, A.; Morris, C.; Blake, D.; Siren, M.; Halliwell, B.; Gustafsson, T.; Loefkvist, B.; Bergelin, I. *Agents Actions* **1990**, *29*, 68-70.
21. Vallejo, M.; Jackson, T.; Lightman, S.; Hanley, M. R. *Nature* **1987**, *330*, 656-658.
22. Nicoletti, F.; Bruno, V.; Cavallaro, S.; Copani, A.; Sortino, M. A.; Canonico, P. L. *Mol. Pharmacol.* **1990**, *37*, 689-693.

23. Willcocks, A. L.; Strupish, J.; Irvine, R. F.; Nahorski, S. R.
    Biochem. J. **1989**, *257*, 297-300.
24. Agranoff, B. W.; Seguin, E.B. *Preparative Biochem.* **1974**, *4*,
    359-366.
25. Hokin, L. E.; Hokin, M. R. *Biochim. Biophys. Acta* **1955**, *18*,
    470-478.
26. Godfrey, P. P. *Biochem. J*. **1989**, *258*, 621-624.
27. Hwang, P. M.; Bredt, D. S.; Snyder, S. H. *Science* **1990**, *249*,
    802-804.
28. Roberts, R. M. *Serendipity. Accidental Discoveries in Science*;
    John Wiley and Son, Inc: New York, NY, **1989**, pp. 198-200.
29. Allison, J. H.; Blisner, M. E.; Holland, W. H.; Hipps, P. P.;
    Sherman, W. R. *Biochem. Biophys. Res. Commun.* **1976**, *68*,
    1332-1338.
30. Low, M. *FASEB J*. **1989**, *3*, 1600-1608.
31. Doering, T. L.; Masterson, W. J.; Hart, G. W.; Englund, P. T.
    J. Biol. Chem. **1990**, *265*, 611-614.
32. Ferguson, M. A. J.; Williams, A. F. *Ann. Rev. Biochem.* **1988**,
    57, 285-320.
33. Lisanti, M. P.; Rodriguez-Boulan, E.; Saltiel, A. R. *Membrane
    Biol.* **1990**, *117*, 1-10.
34. Kennington, A. S.; Hill, C. R.; Craig, J.; Bogardus, C.; Raz, I.;
    Ortmeyer, H. K.; Hansen, B. C.; Romero, G.; Larner, J. *N. Engl.
    J. Med.* **1990**, *323*, 373-378.

RECEIVED February 11, 1991

Chapter 3

Enantiospecific Synthesis and Ca²⁺-Release Activities of Some Inositol Polyphosphates and Their Derivatives

Clinton E. Ballou and Werner Tegge

Department of Molecular and Cell Biology, University of California, Berkeley, CA 94720

D-Myoinositol 1,4,5-trisphosphate, which is released into the cytoplasm of the cell by the action of phospholipase C on 1-phosphatidyl-D-myoinositol 4,5-bisphosphate, acts as a second messenger for mobilization of Ca^{2+} from stores in the endoplasmic reticulum. Enantiospecific syntheses, starting from D- and L-chiroinositol, have been devised for D- and L-myoinositol 1,4,5-trisphosphate, D- and L-myoinositol 1,3,4-trisphosphate, D- and L-chiroinositol 1,3,4-trisphosphate, D- and L-myoinositol 1,3,4,5-tetrakisphosphate, and 1-aminopropylphosphoryl-D-myoinositol 4,5-bisphosphate. The latter was coupled to a resin to provide an affinity matrix and to a photoactivatable crosslinking agent for radiolabeling of specific binding proteins. The compounds differ greatly in their ability to stimulate release of Ca^{2+} by saponin-permeabilized rat basophilic leukemia cells, and the structure-activity relationships provide new insight into the nature of the D-myoinositol 1,4,5-trisphosphate binding site on the endoplasmic reticulum.

This report provides an opportunity to summarize some recent developments in the synthesis of inositol polyphosphates and their derivatives by methods that avoid the necessity for resolution of racemic intermediates or products. Chemical syntheses that take advantage of the inherent chirality of natural starting materials have long had an appeal to organic chemists, particularly in the carbohydrate field (1). Because D- and L-chiroinositol derivatives can be converted in good yield to chiral myoinositol derivatives and because these precursor inositols are available in quantity from plant extracts, the strategy can be exploited with good success (2). As illustrated in this report, the intermediates are such as to provide routes to the synthesis of a variety of useful compounds related to the D-myoinositol 1,4,5-trisphosphate second messenger system.

A Brief Look Backward

The chemistry of inositol phosphates was in its infancy when, in 1959, the structure and stereochemistry of the myoinositol phosphate component of monophosphoinositide (phosphatidyl-myoinositol) were described. Mono- and diphosphates of myoinositol

0097–6156/91/0463–0033$06.00/0

had been isolated from the phospholipids of brain, heart, liver, wheat germ and soya bean (reviewed in *3*). Phytic acid (myoinositol hexaphosphate) was well-characterized and had been hydrolyzed to a monophosphate (*4*), while chemical syntheses of myoinositol 2-phosphate (*5*), pinitol 4-phosphate and L-chiroinositol 3-phosphate (*6*) had been reported.

The general structure of phosphatidyl-myoinositol was recognized at the time, but the linkage to the inositol ring was undefined until it was found that the lipid could be degraded by alkali to yield an optically active myoinositol monophosphate that was readily isomerized by acid to give myoinositol 2-phosphate (*7*). This suggested that the compound was one of the myoinositol 1-phosphate enantiomers, which was confirmed by synthesis of L-myoinositol 1-phosphate, a compound that proved to be the optical antipode of the product from the lipid (*8*). It is of interest that this synthetic isomer also proved to be "natural" when it was later found to be the product of the enzymatic cyclization of D-glucose 6-phosphate (*9*), an intermediate in myoinositol biosynthesis.

Jordi Folch played a major role in unraveling the structures of the brain phospholipids (*10, 11*). A key study (*11*) described the isolation of higher phosphates of phosphatidyl-myoinositol that Folch mistakenly identified as a "diphosphoinositide" because he concluded that the preparation yielded a "myoinositol metadiphosphate" on acid hydrolysis. It is now recognized that this was a mixture of mono-, bis- and trisphosphates resulting from phosphate migration (*12*). Later work (*12, 13*) demonstrated that base-catalyzed degradation of the Folch preparation yielded a defined mixture of two myoinositol monophosphates, two bisphosphates and two trisphosphates, along with glycerol 1- and 2-phosphate. The structures of these products were rationalized as resulting from the degradation of a mixture of 1-phosphatidyl-D-myoinositol, 1-phosphatidyl-D-myoinositol 4-phosphate, and 1-phosphatidyl-D-myoinositol 4,5-bisphosphate (*14*), results that were confirmed by Brown and Stewart (*15*).

Enantiospecific Synthesis of Chiral Inositol Polyphosphates

Several methods have been described for synthesis of myoinositol polyphosphates, most of which begin with the readily available myoinositol (*16*). Because myoinositol is a *meso* compound, however, the intermediates are racemic mixtures and chemical or biochemical resolution is required to give the pure enantiomers. Although clever and effective methods have been developed for achieving such resolution, they introduce additional steps and may leave uncertainty regarding the optical purity of the product.

The strategy we have adopted for synthesis of inositol polyphosphates takes advantage of the inherent chirality of D- and L-chiroinositol. These compounds are available from D-pinitol and L-quebrachitol, methyl ethers that are converted to free inositol by treatment with HI. Pinitol reaches 20% of the dry weight of the cold water extract of sugar pine (*Pinus lambertiana Dougl*) stump wood (*17*), while quebrachitol comes from the rubber tree (*Hevea brasiliensis*) (*18*). The initial steps in our syntheses take advantage of the known reaction of chiral inositols to yield the 1,2:5,6-dicyclohexylidene derivative (*19*), which can be benzylated and then hydrolyzed in acid to give the 3,4-dibenzyl-chiroinositol (Figure 1) (*20*). Because the equatorial hydroxyl groups are more readily benzylated than the axial ones, and because the compound has an axis of symmetry, 1,2,5-tribenzoyl-3,4-dibenzyl-chiroinositol can be prepared in good yield and is the product obtained regardless of which axial hydroxyl group is substituted (Figure 2). Finally, inversion of the free hydroxyl group, via the trifluoromethanesulfonate ester, yields chiral myoinositol derivatives (Figure 3). In this reaction, the presumed benzoyloxonium ion opens preferentially to give the

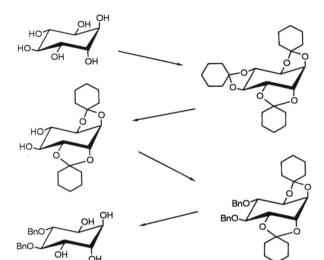

Figure 1. Conversion of D-chiroinositol, via the 1,2:5,6-dicyclohexylidene derivative, to 3,4-dibenzyl-D-chiroinositol. Bn, benzyl.

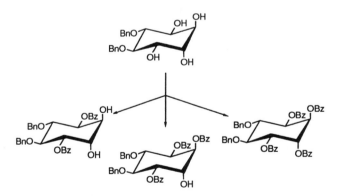

Figure 2. Partial benzoylation of 3,4-dibenzyl-D-chiroinositol, with 3 equivalents of benzoyl chloride, to yield 1,2,5-tribenzoyl-3,4-dibenzyl-D-chiroinositol as the major product. Bz, benzoyl.

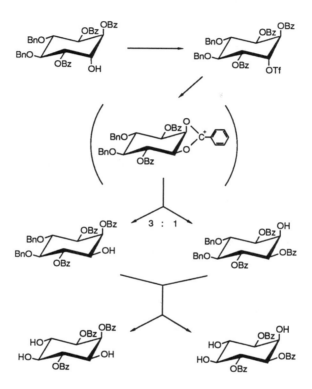

Figure 3.   Conversion of 1,2,5-tribenzoyl-3,4-dibenzyl-D-chiroinositol, by displacement of the trifluoromethanesulfonate ester, to a 3:1 mixture of tribenzoyl-dibenzyl-D-myoinositols and their hydrogenolysis to yield 2,3,6- and 1,3,6-tribenzoyl-D-myoinositol. Tf, trifluoromethanesulfonyl.

product with an axial benzoyl ester, 2,3,6-tribenzoyl-4,5-dibenzyl-myoinositol, the key intermediate for synthesis of D- and L-myoinositol 1,4,5-trisphosphate (2). The tribenzoyl-myoinositols for these syntheses are obtained by hydrogenolysis of the benzyl ether groups (Figure 3). The phosphorylation, by a dibenzylphosphite procedure (21), and the unblocking reactions for synthesis of D-myoinositol 1,4,5-trisphosphate follow the steps in Figure 4. By similar reactions, D- and L-myoinositol 2,4,5-trisphosphate are available from the minor product of the inversion reaction, whereas D- and L-chiroinositol 1,3,4-trisphosphate have been prepared in parallel reactions starting from the 1,2,5-tribenzoyl-3,4-dibenzyl-chiroinositols (Tegge, W.; Ballou, C.E., unpublished data).

The mixture of tribenzoyl-dibenzyl-myoinositol isomers obtained in the inversion reaction (Figure 3) can be substituted with the acid-labile and alkali-stable methoxyethoxymethyl group, the isomers separated by chromatography, and the 2-methoxyethoxymethyl-4,5-dibenzyl-myoinositol derivative phosphorylated and unblocked to yield myoinositol 1,3,4-trisphosphate (Figure 5). In this reaction series, a D-myoinositol intermediate gives an L-myoinositol trisphosphate. By a more extended series of reactions, D-myoinositol 1,3,4,5-tetrakisphosphate is available. This procedure involves debenzoylation of the products of the inversion reaction, followed by partial benzoylation of the 4,5-dibenzyl-myoinositol with two equivalents of benzoyl chloride, during which the major product is 1,3-dibenzoyl-4,5-dibenzyl-myoinositol (Figure 6) (Tegge, W.; Ballou, C. E., unpublished data). Substitution of the two free hydroxyls with the methoxyethoxymethyl group, followed by debenzylation and debenzoylation, provides a 2,4-disubstituted myoinositol from which the tetrakisphosphate is available.

Synthesis of Affinity Ligands and Crosslinking Reagents

We have exploited a general methodology for attaching an extender arm to the phosphate at position 1 of myoinositol 1,4,5-trisphosphate, and then we have utilized this group for attachment of the ligand to an insoluble matrix or to a crosslinking agent. Jina and Ballou (22) have described the synthesis of such derivatives based on *trans*-1,2-cyclohexanediol bisphosphate as an analog of myoinositol 4,5-bisphosphate. We have now applied a similar strategy (Figure 7) for synthesis of the myoinositol trisphosphate derivative. The tribenzoyl-dibenzyl-myoinositols from the inversion reaction are substituted with the tetrahydropyranyl group, the isomers separated by chromatography, and the appropriate isomer is debenzylated and phosphorylated. After removal of the tetrahydropyranyl group and reaction with O-(N-carbobenzyloxy-3-aminopropyl)-O-benzyl-di-N-iso-propylphosphoramidite, a derivative is obtained that, after unblocking, yields 1-(3-aminopropylphosphoryl)-D-myoinositol 4,5-bisphosphate. This product has been coupled to carbonyldiimidazole-activated agarose and to 4-azidosalicylic acid to provide derivatives for use in study of specific binding protein (Denis, G. V.; Tegge, W.; Ballou, C. E., unpublished data).

Ca²⁺-Release Activities of Inositol Polyphosphates and Derivatives

One measure of the biological activity of myoinositol trisphosphate is its ability to stimulate release of Ca²⁺ from endoplasmic reticulum stores of saponin-permeabilized rat basophilic leukemia cells (23). We used this system to assess the activity of our synthetic derivatives and to answer three questions: are the compounds prepared by our enantiospecific strategy similar in biological activity to preparations of the same compounds obtained by others; do the new derivatives we have made increase our understanding of the structural requirements for stimulation of Ca²⁺ release; and do the

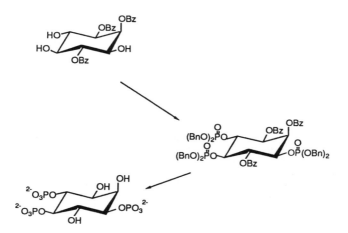

Figure 4. Synthesis of D-myoinositol 1,4,5-trisphosphate from 2,3,6-tribenzoyl-D-myoinositol.

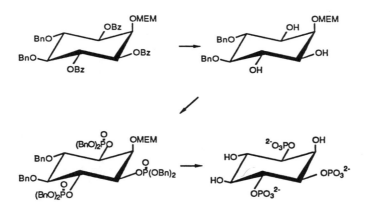

Figure 5. Synthesis of L-myoinositol 1,3,4-trisphosphate from 1,3,6-tribenzoyl-4,5-dibenzyl-D-myoinositol. MEM, methoxyethoxymethyl.

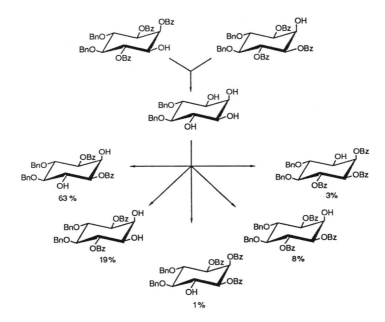

Figure 6. Partial benzoylation of 4,5-dibenzyl-D-myoinositol with 2 equivalents of benzoyl chloride to give 1,3-dibenzoyl-4,5-dibenzyl-D-myoinositol as the major product.

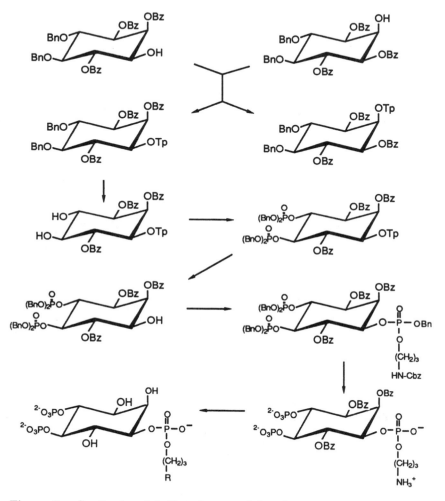

Figure 7.   Synthesis of 1-(3-aminopropylphosphoryl)-D-myoinositol 4,5-bisphosphate. Tp, tetrahydropyranyl; R, amino or amide.

affinity ligands we have prepared retain sufficient biological activity to warrant their use as general probes for study of myoinositol trisphosphate binding proteins?

With regard to the first question, our synthetic D-myoinositol 1,4,5-trisphosphate is similar in activity to that reported for other preparations (24), but our preparation of L-myoinositol 1,4,5-trisphosphate shows no activity while others report a low but measureable activity (Table I). Our D-myoinositol 2,4,5-trisphosphate is also less active than reported for other preparations, possibly because it contains less of the 1-4-5-trisphosphate isomer.

Table I. Ca²⁺-Release Activities of Polyol Phosphates

Compound	EC$_{50}$ [μM][a]
D-myoinositol 1,4,5-trisphosphate	0.17[b]
iodoazidosalicylaminopropyl derivative	0.26
aminopropyl derivative	6.3
D-myoinositol 2,4,5-trisphosphate	4.3
D-chiroinositol 1,3,4-trisphosphate	4.2
L-myoinositol 2,4,5-trisphosphate	110
L-chiroinositol 1,3,4-trisphosphate	120
L-myoinositol 1,4,5-trisphosphate	>2000

[a]Concentration for half-maximal stimulation

[b]Values from 0.1 to 1 μM have been reported (24).

Concerning the second question, we find that D-chiroinositol 1,3,4-trisphosphate and D-myoinositol 2,4,5-trisphosphate both show about 1/25th the activity of D-myoinositol 1,4,5-trisphosphate. The D-chiroinositol trisphosphate can be considered as an analog of D-myoinositol 1,4,5-trisphosphate with an axial phosphate at position 1 or as an analog of D-myoinositol 2,4,5-trisphosphate with an axial hydroxyl at position 1. More surprising are the relatively high activities of L-myoinositol 2,4,5-trisphosphate and L-chiroinositol 1,3,4-trisphosphate, which have about 1/700th the activity of D-myoinositol 1,4,5-trisphosphate and are clearly more active than the L-isomer of the latter. We postulate that the axial phosphate group in these compounds could lead to a more flexible inositol ring conformation and, thereby, allow the bisphosphate groups to adapt to the active site of a binding protein.

Finally, the D-myoinositol 1,4,5-trisphosphate derivatives with an extender arm do show significant Ca²⁺-release activities (Table I) and the iodinated azidosalicyl derivative has an activity that is essentially equal to that of the parent myoinositol trisphosphate. The only difference we note in the two is that the stimulation of release by the crosslinking reagent is more rapidly dissipated than that by the D-myoinositol 1,4,5-trisphosphate. This result could reflect a more rapid metabolism of the former or association of the derivative with membrane lipids or hydrophobic membrane proteins.

Literature Cited

1. *Synthetic Methods for Carbohydrates*; El Khadem, H. S., Ed.; ACS Symposium Series 39; American Chemical Society: Washington, D. C. 1976.
2. Tegge, W.; Ballou, C. E. *Proc. Natl. Acad. Sci. U.S.A.* **1989**, *86*, 94.
3. Angyal, S. J.; Anderson, L. *Adv. Carbohydr. Chem.* **1959**, *14*, 135.
4. McCormick, M. H.; Carter, H. E. *Biochem. Preps.* **1952**, *2*, 65.
5. Iselin, B. *J. Am. Chem. Soc.* **1949**, *71*, 3822.
6. Kilgour, G. L.; Ballou, C. E. *J. Am. Chem. Soc.* **1958**, *80*, 3956.
7. Pizer, F. L.; Ballou, C. E. *J. Am. Chem. Soc.* **1959**, *81*, 915.
8. Ballou, C. E.; Pizer, L. I. *J. Am. Chem. Soc.* **1960**, *82*, 3333.
9. Eisenberg, F. *J. Biol. Chem.* **1967**, *242*, 1375.
10. Folch, J. *J. Biol. Chem.* **1949**, *177*, 497.
11. Folch, J. *J. Biol. Chem.* **1949**, *177*, 505.
12. Grado, C.; Ballou, C. E. *J. Biol. Chem.* **1961**, *236*, 54.
13. Tomlinson, R. V.; Ballou, C. E. *J. Biol. Chem.* **1961**, *236*, 1902.
14. Brockerhoff, H.; Ballou, C. E. *J. Biol. Chem.* **1961**, *236*, 1907.
15. Brown, D. M.; Stewart, J. C. *Biochim. Biophys. Acta* **1966**, *125*, 413.
16. Billington, D. C. *Chem. Soc. Rev.* **1989**, *18*, 83.
17. Anderson, A. B. *Ind. Eng. Chem.* **1953**, *45*, 593.
18. van Alphen, J. *Ind. Eng. Chem.* **1951**, *43*, 141.
19. Jiang, D.; Baker, D. C. *J. Carbohydr. Chem.* **1986**, *5*, 615.
20. Angyal, S. J.; Stewart, T. S. *Aust. J. Chem.* **1966**, *19*, 1683.
21. Yu, K.-L.; Fraser-Reid, B. *Tetrahedron Lett.* **1986**, *29*, 979.
22. Jina, A. N.; Ballou, C. E. *Biochemistry* **1990**, *29*, 5203.
23. Meyer, T.; Wenzel, T.; Stryer, L. *Biochemistry* **1990**, *29*, 32.
24. Berridge, J. J.; Irvine, R. F. *Nature* **1984**, *312*, 315.

RECEIVED February 11, 1991

Chapter 4

Synthesis of Inositol Polyphosphates and Their Derivatives

Shoichiro Ozaki and Yutaka Watanabe

Department of Resources Chemistry, Faculty of Engineering, Ehime University, Matsuyama, Japan 790

D-*myo*-Inositol 1,4,5-trisphosphate [D-Ins(1,4,5)P_3], D-Ins(2,4,5)P_3, D-Ins(1,3,4)P_3, D-Ins(1,3,4,5)P_4, Ins(1,3,4,6)P_4, and 2-deoxy-Ins(1,4,5)P_3 were synthesized from *myo*-inositol. *o*-Xylylene *N,N*-diethylphosphoramidite (OXDEP) was found to be an efficient phosphitylating agent. Methyl hydrogen 2,3-O-cyclohexylidenetartrate (MHCT) was found to be useful for the enantioselective acylation of inositol derivatives yielding the corresponding monotartrates with high optical purities. These new methods permitted the simple and efficient preparation of D-Ins(1,4,5)P_3 and D-Ins(1,3,4,5)P_4 as well as other InsP_x derivatives. InsP_x affinity columns were obtained.

Recent biological studies have revealed that D-*myo*-inositol 1,4,5-trisphosphate (Ins(1,4,5)P_3) acts as a cellular second messenger leading to intense biological interest in this substance.[1] This material was isolated in 1961 by C. E. Ballou by chemical hydrolysis of bovine brain phosphoinositide,[2] but the isolation process was very difficult and the large quantities required for biological studies are simply not available by this method. In 1986, we reported the first total synthesis of Ins(1,4,5)P_3.[3] Since then we have been studying the synthesis of its metabolites and derivatives, focusing on the development of easy and simple synthetic methods. Three important problems must be overcome in order to get desired products efficiently: selective protection and deprotection, optical resolution, and phosphorylation.[4]

Synthesis of D-*myo*-Inositol 1,4,5-Trisphosphate (1).[3] Treatment of *myo*-inositol (2) with 1-ethoxycyclohexene yielded 1,2:4,5-biscyclohexylidene derivative **3** which was easily isolated by crystallization (see Figure 1). The remaining hydroxyls were benzylated with benzyl chloride and sodium hydride followed by removal of the less stable 4,5-cyclohexylidene group with ethylene glycol and TsOH to give the 4,5-diol **5**. Bisallylation followed by cleavage of the remaining cyclohexylidene group with 80% aqueous acetic acid solution at 80 °C afforded 4,5-diallyl-3,6-dibenzyl-*myo*-inositol (**6**). This racemic diol was subjected to optical resolution in the next step.

After a number of trials, we prepared the diastereomeric mixture of 1-*l*-menthoxyacetyl derivatives **7a** and **7b**, from which the desired product was easily obtained by direct crystallization. Alkaline hydrolysis of the ester gave the enantiomerically pure 1,2-diol **6a**. Selective allylation of the equatorial hydrogen group at C-1 proceeded easily with allyl bromide and NaOH in benzene. After benzylation of the remaining hydroxyl, deallylation was achieved by isomerization with tris(triphenylphosphine) rhodium chloride. Subsequent acidic hydrolysis afforded the key synthetic intermediate, D-2,3,6-tri-O-benzyl-*myo*-inositol (**10**). The triol **10** was perphosphorylated with dianilinophosphoric chloride (DAPC). Deprotection of the fully protected derivative thus obtained was accomplished by treatment with *iso*-amyl nitrite, acetic acid, and acetic anhydride in pyridine followed by hydrogenolysis over

0097–6156/91/0463–0043$06.75/0

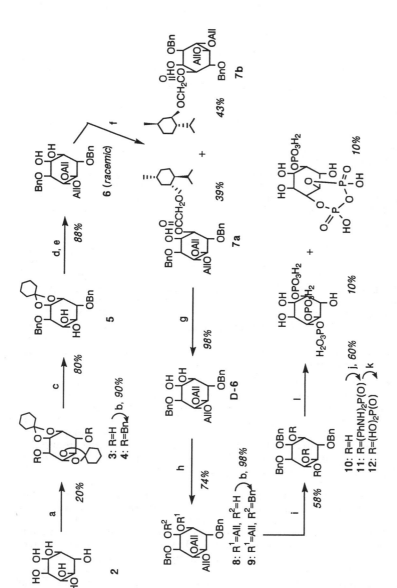

Figure 1. Synthesis of D-*myo*-inositol 1,4,5-trisphosphate.

a: 1-EtO-cyclohexene/TsOH, b: NaH/BnCl/DMF, c: (HOCH₂)₂/TsOH, d: NaH/AllBr/DMF, e: aq. AcOH, f: *l*-MntAcCl/Py, g: aq. NaOH/MeOH, h: AllBr/NaOH, i: RhCl(PPh₃)₃/DABCO then HCl/MeOH, j: (PhNH)₂P(O)Cl/Py, k: *i*-AmONO/Ac₂O/AcOH/Py, l: H₂/5% Pd-C

Pd/C to give the desired product, Ins(1,4,5)P$_3$ (**1**). 1-Phospho-4,5-pyrophospho-*myo*-inositol (**13**) was also obtained as a by-product.

Synthesis of Ins(2,4,5)P$_3$ (14).[5] Efforts to phosphorylate 1,3,6-tribenzyl-*myo*-inositol by DAPC were unsuccessful mainly because the hydroxyl group at the 2 position (axial hydroxyl) is unreactive. However, the axial hydroxyl group of **17** did react with phosphoric trichloride followed by benzyl alcohol and the resulting phosphite was oxidized with *t*-butyl hydrogen peroxide (PCl$_3$ method) to afford **18** (see Figure 2).

3,6-Dibenzyl-1,2-cyclohexylidene-*myo*-inositol (**5**) was *p*-methoxybenzylated at the 4,5 positions to afford 3,4-dibenzyl-1,2-cyclohexylidene-4,5-di(*p*-methoxybenzyl)-*myo*-inositol (**15**). The 1,2-cyclohexylidene group in **15** was cleaved by hydrochloric acid to give diol **16**. The 1-hydroxyl group in **16** was selectively menthoxyacetylated giving diastereomers **17**, which were easily separated by chromatography on silica gel. The PCl$_3$ method was applied to the 2-hydroxyl group in **17** to give **18**. Methoxybenzyl groups were cleaved by DDQ to give diol **19**. One hydroxyl group in **19** was esterified with levulinic acid and the remaining hydroxyl was phosphorylated by PCl$_3$ method to give **21**. After cleavage of levulinyl group in **21** by hydrazine, the resulting hydroxyl group in **22** was phosphorylated to give trisphosphate **23**. The PCl$_3$ method is not preferred for phosphorylation of vicinal diols. Tetrabenzyl pyrophosphate (TBPP) was found to be effective for the phosphorylation of vicinal diols.[6]

Synthesis of Ins(1,3,4)P$_3$ (24).[7] The preparation of this compound is shown in Figure 3. The 5,6-hydroxyl groups of 1,2:3,4-bis(cyclohexylidene)-*myo*-inositol (**25**) were benzylated. The 3,4-cyclohexylidene group in **26** was cleaved selectively and the generated hydroxyl groups at C-3 and C-4 were *p*-methoxybenzylated. The 1,2-cyclohexylidene functionality was cleaved. The 1-hydroxyl group in **29** was reacted with menthoxyacetyl chloride and the diastereomeric mixture was separated.

The menthoxyacetyl group in **30** was cleaved and the 1-hydroxyl group was methoxymethylated with triethylmethoxymethylammonium chloride. After benzylation of the 2-hydroxyl group in **31**, the *p*-methoxybenzyls were cleaved by DDQ and the methoxymethyl group was cleaved by hydrochloric acid. D-2,5,6-Tri-O-benzyl-*myo*-inositol (**33**) was phosphorylated with TBPP. Hydrogenolysis of **34** afforded optically active Ins(1,3,4)P$_3$ (**24**).

Synthesis of Ins(1,3,4,5)P$_4$ (35).[8] 1,2:4,5-Bis(cyclohexylidene)-*myo*-inositol was benzoylated at the 3-position with benzoyl imidazolide in the presence of CsF (benzoyl chloride and benzoic anhydride showed lesser selectivity; see Figure 4). The 6-hydroxyl group in **36** was benzylated with benzyl trichloroacetimidate in the presence of trifluoromethanesulfonic acid. The 4,5-cyclohexylidene group in **37** was cleaved and then the 4,5-hydroxyl group was benzoylated. The 1,2-cyclohexylidene was cleaved and the racemic diol was menthoxyacetylated and separated using medium pressure liquid chromatography. The 2-hydroxyl in **40** was benzylated, the benzoates were cleaved, the free hydroxyls in **41** were phosphorylated, and then the protecting groups removed to give the desired compound **35**.

Synthesis of *myo*-Inositols from Glucuronolactone.[9] *myo*-Inositol derivatives may be obtained from glucuronolactone **43** (see Figure 5). In this case optical resolution is unnecessary, but epimerization at C-5 (glucose numbering) *gluco*-to *ido*-configuration is required. Conversion of **43** to the 1,2-acetonide derivative was followed by tosylation of the remaining free hydroxyl group. The lactone **44** was

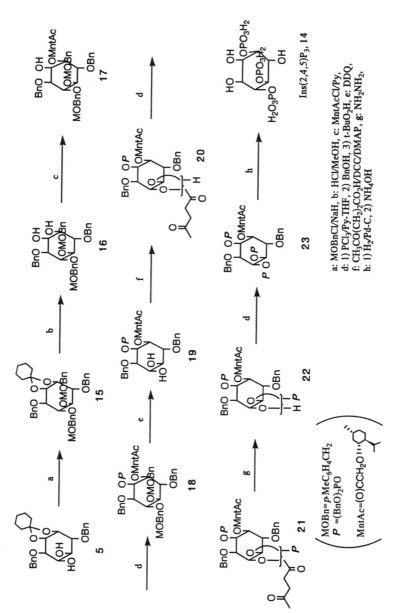

Figure 2. Synthesis of D-*myo*-inositol 2,4,5-trisphosphate.

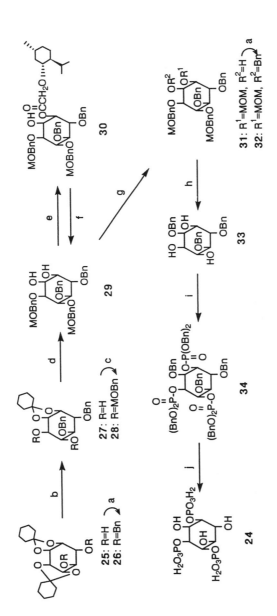

a: NaH/BnCl/DMF, b: (CH₂OH)₂/TsOH/CHCl₃, c: NaH/MOBnCl/DMF, d: HCl/MeOH, e: MntAcCl/Py,
f: NH₃/MeOH, g: 1) n-Bu₂SnO; 2) MOM-NEt₃ Cl, h: 1) DDQ; 2) HCl, i: n-BuLi/TBPP, j: H₂/5% Pd-C

Figure 3. Synthesis of D-*myo*-inositol 1,3,4-trisphosphate.

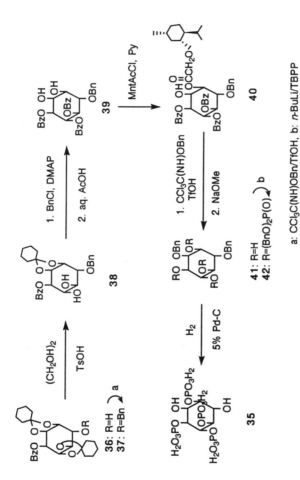

Figure 4. Synthesis of D-*myo*-inositol 1,3,4,5-tetrakisphosphate.

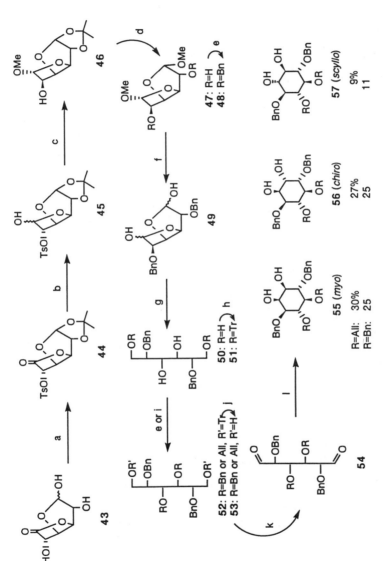

Figure 5. Synthesis of optically active inositols from glucuronolactone.

a: 1) acetone/TsOH [64%]; 2) TsCl/Py [88%], b: *i*-Bu₂AlH [88%], c: K₂CO₃/MeOH [87%], d: MeOH/HCl [100%], e: NaH/BnCl [73% for **48**], f: aq. H₂SO₄ [91%], g: NaBH₄[75%], h: TrCl/Py [80%], i: NaH/AllBr[93%], j: aq. HCl [50% for **53** (R=Bn) from **51**, 91% for **53** (R=All)], k: (COCl)₂/DMSO/Et₃N, l: TiCl₄/Zn(Cu)

reduced with *iso*-butylaluminium hydride to the lactol **45**, which was treated with potassium carbonate and methanol to invert the configuration at C-5 through an oxirane intermediate. Acidic methanolysis of **46** followed by benzylation, hydrolysis, and reduction afforded 2,5-di-O-benzyl-iditol **50**. Tritylation with trityl chloride, allylation with allyl bromide, and hydrolysis then gave the 1,6-diol **53**. Oxidation of the diol according to Swern gave the dialdehyde **54**. Cyclization of this dialdehyde using a low valent titanium species gave *myo*-, *chiro*-, and *scyllo*-inositol derivatives **55**, **56**, and **57** in 30, 27, and 9% yields respectively. In order to get the *myo*-configuration exclusively, the diols can be converted to the olefin with triphenylphosphine and 2,4,5-triiodoimidazole. The olefin is then oxidized with OsO_4 to afford the *myo*-inositol derivative in 75% overall yield from the diol mixture.

Phosphorylation Methods. See Table 1 for a list of some commonly-employed phosphorylating reagents. DAPC was used for the first synthesis of Ins(1,4,5)P$_3$, but this reagent is not effective for the synthesis of Ins(2,4,5)P$_3$. The PCl$_3$ method is effective for the phosphorylation of an isolated hydroxyl group, even at the hindered 2-position, but not for vicinal diols. TBPP on the other hand, has been employed for the phosphorylation of polyols, but, in some instances, the yield is not satisfactory. Because of the limitation of the existing phosphorylating reagents, we explored the utility of *o*-xylylene diethylphosphoramidite (OXDEP, **58**),[10] which is easily obtainable from *o*-xylylenediol and hexaethylphosphoroustriamide (**60**) (see Figure 6). OXDEP reacts readily with alcohols in the presence of tetrazole to give the trivalent phosphite species **61** which may be then oxidized either with mCPBA to give the phosphate **62** or with sulfur to give the thiophosphate **63**. Finally, the *o*-xylylene group undergoes facile hydrogenolysis to give the free phosphate **64**. This phosphorylation process is easy and gives high yields.

Synthesis of Ins(1,3,4,6)P$_4$ (66).[11] The 3,5-dihydroxy groups of 1,6:3,4-bis(tetraisopropylidene) disilyloxy-*myo*-inositol (**67**) were benzoylated and the TIPS protecting groups in **68** were cleaved by hydrogen fluoride (see Figure 7). The resulting tetrol **69** was phosphorylated by OXDEP method. Hydrogenolysis of **70** with hydrogen and Pd/C followed by treatment with sodium methoxide gave Ins(1,3,4,6)P$_4$ (**66**).

Synthesis of 2-deoxy Ins(1,4,5)P$_3$ (71). The synthesis of 2-deoxy Ins(1,4,5)P$_3$ (**71**) proceeded from 4,5-dibenzoyl-3,6-dibenzyl-1,2-cyclohexylidene-*myo*-inositol (**72**), whose synthesis is described above (see Figure 8). The cyclohexylidene group was cleaved by 80% aqueous acetic acid to give the 1,2-diol **73**. Treatment of the dibutylstannylene derivative of the diol with methoxymethyl chloride regioselectively afforded the C-1 monoprotected MOM ether **74**. A four step sequence was employed to deoxygenate at the 2 position. First, the free hydroxyl group was oxidized with PCC to the ketone **75**, which was smoothly converted to the tosyl hydrazone **76**. Then hydrazone was reduced with sodium cyanoborohydride, and the resulting hydrazine derivative **77** was cleaved with sodium acetate to give 2-deoxy compound **78**. Removal of the MOM and benzoyl protecting groups gave the triol **80**, which was phosphorylated with TBPP and butyllithium. Hydrogenolysis of **81** gave 2-deoxy Ins(1,4,5)P$_3$ (**71**).

Simple Synthesis of Ins(1,3,4,5)P$_4$ (35) and Ins(1,4,5)P$_3$ (1).[12] When *myo*-inositol was reacted with 3.5 molar equivalent of benzoyl chloride, a 37% yield of 1,3,4,5-tetrabenzoyl-*myo*-inositol (**82**) was obtained along with lesser yields of seven other partially protected derivatives as shown in Table 2. From the major product, an important metabolite, Ins(1,3,4,5)P$_4$ (**35**) was obtained as follows (Figure

Table 1. Phosphorylating methods

1. DAPC method

1) 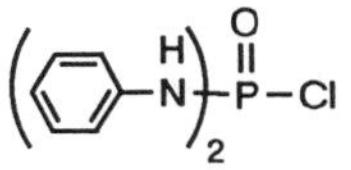 $\left(\underset{H}{\underset{|}{\bigcirc}}\text{-N} \right)_2 \overset{O}{\underset{\|}{P}}\text{-Cl}$ 2) $i\text{-C}_5\text{H}_{11}\text{ONO, AcOH}$

2. PCl$_3$ method

1) PCl$_3$ 2) BnOH   3) t-BuOOH

3. TBPP method

1) BuLi   2) $\left(\text{BnO}\right)_2 \overset{O}{\underset{\|}{P}}\text{-O-}\overset{O}{\underset{\|}{P}}\left(\text{OBn}\right)_2$

4. OXDEP method

1) 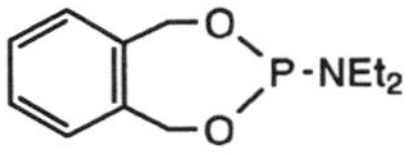 [structure] P-NEt$_2$ +   Tetrazole   2) mCPBA or S$_8$

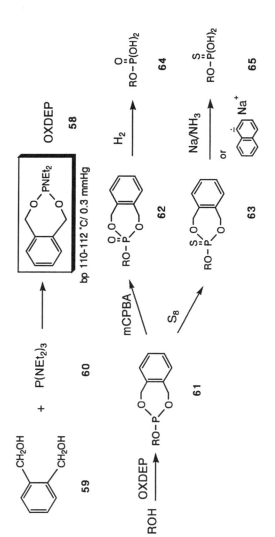

Figure 6. *O*-Xylylene *N,N*-Diethylphosphoroamidite (OXDEP).

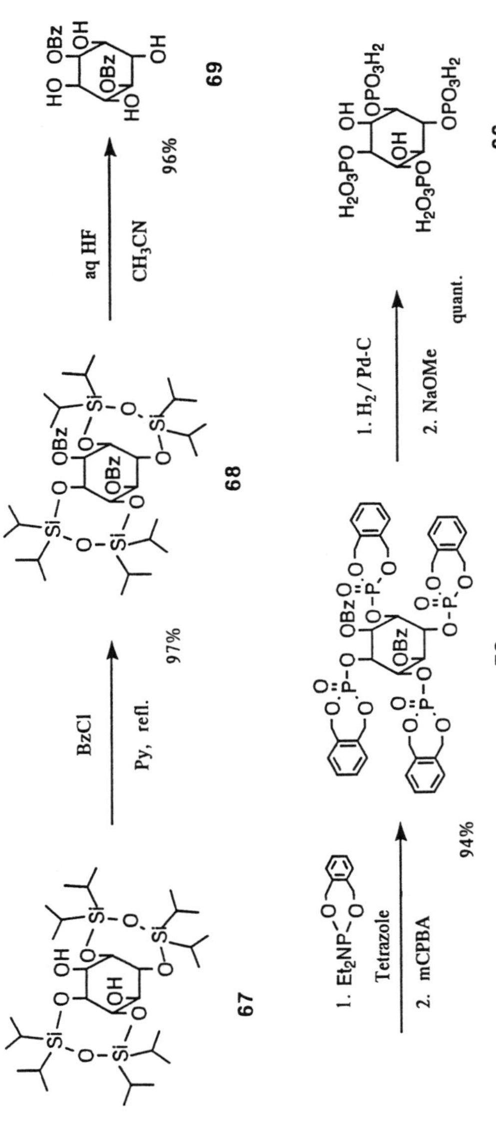

Figure 7. Synthesis of *myo*-inositol 1,3,4,6-tetrakisphosphate.

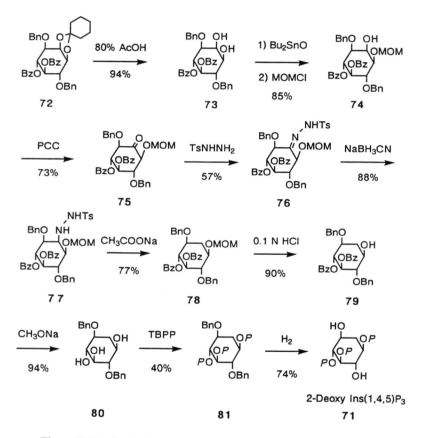

Figure 8. Synthesis of 2-deoxy-*myo*-inositol 1,4,5-trisphosphate.

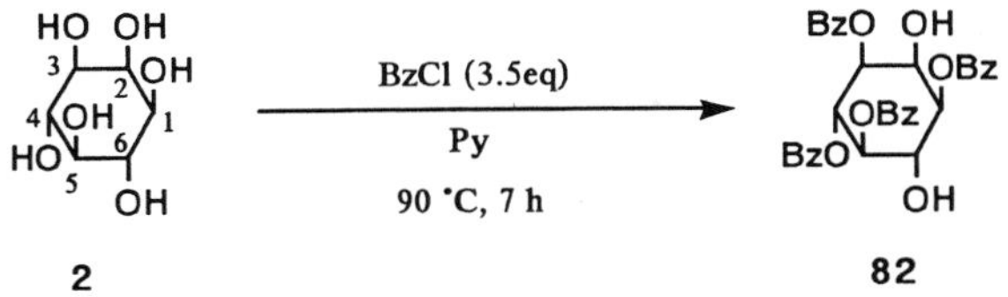

InsBzx	Yield /%
Ins(1,3,4,5)Bz$_4$	37
Ins(1,3,5)Bz$_3$	15
Ins(1,3,4,5,6)Bz$_5$	14
Ins(1,3,4,6)Bz$_4$	13
Ins(1,3,4)Bz$_3$	9
Ins(1,4,5)Bz$_3$	6
Ins(1,5,6)Bz$_3$	1
Ins(1,5)Bz$_2$	1
Total	96

Table 2. Benzoylation of *myo*-inositol

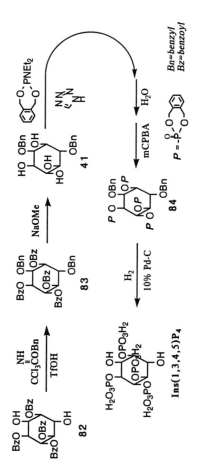

Figure 9. Simple synthesis of *myo*-inositol 1,3,4,5-tetrakisphosphate.

9). First, 1,3,4,5-tetrabenzoyl inositol (**82**) was benzylated with benzyl trichloroacetimidate in the presence of trifluoromethanesulfonic acid. After cleavage of the benzoyl groups with sodium methoxide, the resulting tetraol **41** was phosphorylated with OXDEP. Hydrogenolysis then provided $Ins(1,3,4,5)P_4$. Alternatively, the 2,6-positions of $Ins(1,3,4,5)$tetrabenzoate may be protected with *t*-butyldimethylsilyl groups. Reaction of the resulting product **85** with Grignard reagents under various conditions gave 1,3,4,5-tetrol **86**, 1,4,5-triol **87**, and 1,4-diol **88** in varying ratios (Figure 10). The ratio appears to depend upon the size of the halogen atom of the Grignard reagent as well as on the choice of solvent. Under the optimal conditions, 53% of the 1,4,5-triol **87** was obtained, from which $D,L-Ins(1,4,5)P_3$ was easily prepared.

Enantioselective acylation with methyl hydrogen 2,3-O-Cyclohexylidene-Tartrate.[13] One of the minor products isolated in the direct benzoylation of *myo*-inositol was the *meso*-isomer, 1,3,5-tribenzoyl-*myo*-inositol (**89**). It occurred to us that selective acylation of one of the enantiotopic hydroxylic groups at the 4- and 6-positions would result in the formation of an optically active *myo*-inositol derivative *without need for a resolution*. Towards this end, we evaluated many many chiral, non-racemic acylating agents, including menthoxyacetyl chloride, menthoxy carbonyl chloride, amino acid mixed anhydrides, and isocyanates, but were unable to obtain useful selectivities. In contrast, the methanesulfonyl mixed anhydride of certain tartrate mono-ester derivatives, most notably methyl hydrogen 2,3-O-cyclohexylidene tartarate (MHCT), were remarkably effective in the enantioselective acylation process. As shown in Figure 11, a 62% yield of the D-1,3,4,5-tetraacylated *myo*-inositol derivative **90D** with a diastereomeric excess of 96% was obtained by treatment of the *meso*-1,3,5-tribenzoyl *myo*-inositol (**89**) with mixed anhydride of MHCT. If recovery of starting material (28%) is considered, the chemical yield is 86%. Conversion to $D-Ins(1,3,4,5)P_4$ was easily achieved by conventional methods as shown in Figure 11.

MHCT is also useful for the kinetic resolution of racemic 1,3,4,5-tetrabenzoyl-*myo*-inositol (**82**), with the L-isomer undergoing rapid acylation at the 6-position. The results are shown in Figure 12, and indicate that, as is typically observed in kinetic resolutions the enantiomeric excess of the unreacted starting material increases as the conversion is increased. Thus, it was possible to obtain a 41% yield (80% of the theoretical) of 95% ee $D-Ins(1,3,4,5)P_4$ after kinetic resolution with MHCT.

Simple Synthesis of $D-Ins(1,4,5)P_3$ and $D-Ins(1,3,4,5)P_4$. Reagents and techniques mentioned above are combined as shown in Figure 13. 1,3,5-Tribenzoyl-4-tartaroyl-*myo*-inositol (**90D**) was reacted with triethylsilyl chloride to give 2,6-bis(triethylsilyl) derivative **92**. The reaction of **92** with ethylmagnesium bromide afforded 3-benzoyl-2,6-bis(triethylsilyl)-*myo*-inositol (**93**) and 2,6-bis(triethylsilyl)-*myo*-inositol (**94**). When these were phosphorylated and then hydrogenolized, **93** gave $D-Ins(1,4,5)P_3$ and **94** gave $D-Ins(1,3,4,5)P_4$.

Synthesis of $Ins(1,4,5)P_3$ derivatives. 4,5-Bis(dibenzyl phosphoryl)-3,4-diben-zyl-*myo*-inositol **97** is a useful intermediate as shown in Figure 14.[14] The 1-hydroxyl group was silylated and the 2-hydroxyl was *p*-nitrobenzoylated. The 1-triethylsilyl group in **99** was cleaved and that position phosphorylated with PCl_3, BnOH, and *t*-butyl hydrogen peroxide. All benzyl and nitro groups in **101** were reduced with H_2 and Pd/C to give **102**. The aromatic ring in **102** was reduced with hydrogen in the presence of ruthenium oxide to give **104** as shown in Figure 15. The amine **102** was converted to a diazonium salt and the reaction of the salt with sodium

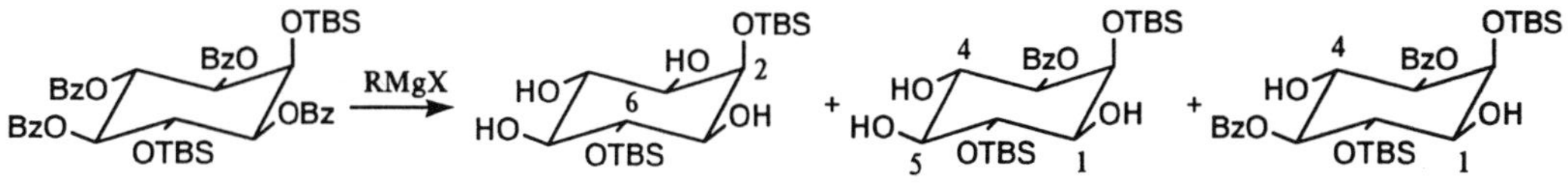

85 **86** **87** **88**

RMgX	eq.	solv.	temp.	time/min	yield/% 86	87	88
MeMgI	40	THF	reflux	180	13	24	32
〃	35	Et_2O	r.t.	30	3	24	56
〃	〃	〃	reflux	40	17	53	12
EtMgI	〃	〃	〃	30	33	36	-
EtMgBr	〃	THF	r.t.	120	25	8	-
〃	〃	Et_2O	reflux	30	51	19	-
〃	〃	〃	〃	160	71	6	-

Figure 10. Reaction of 2,6-bis(*t*-butyldimethylsilyl)-1,3,4,5-tetrabenzoyl-*myo*-inositol with Grignard reagent.

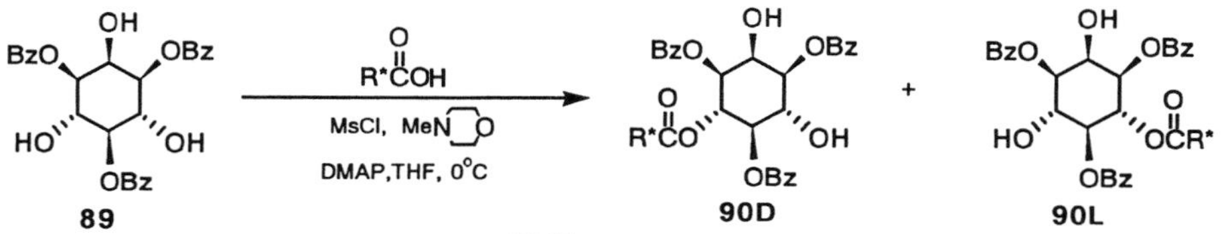

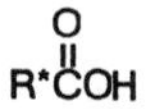 R*COH	Yield Prdct.	SM	Ratio (D : L)	O.P. (de %)
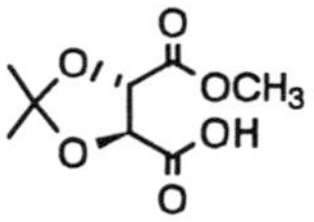 (acetonide, OCH₃, OH)	64%	30%	96 : 4	92
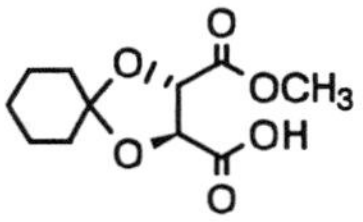 (cyclohexylidene, OCH₃, OH)	62%	28%	98 : 2	96
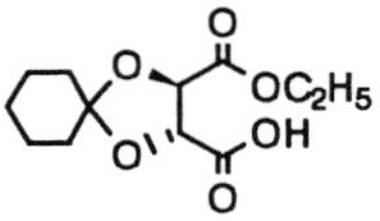 (cyclohexylidene, OC₂H₅, OH)	42%	21%	2 : 98	96
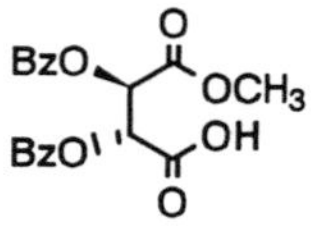 (BzO, BzO, OCH₃, OH)	40%	35%	27 : 73	46

Figure 11. Enantioselective acylation of 1,3,5-tribenzoyl-*myo*-inositol.

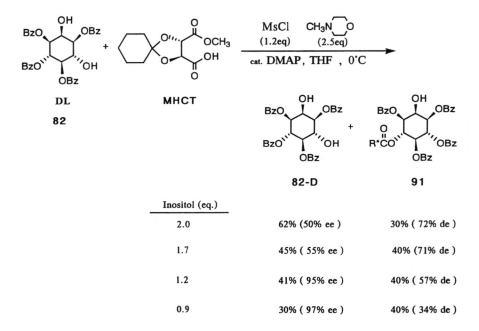

Inositol (eq.)	82-D	91
2.0	62% (50% ee)	30% (72% de)
1.7	45% (55% ee)	40% (71% de)
1.2	41% (95% ee)	40% (57% de)
0.9	30% (97% ee)	40% (34% de)

Figure 12. Kinetic resolution of D,L-1,3,4,5-tetrabenzoyl-*myo*-inositol by OXDEP.

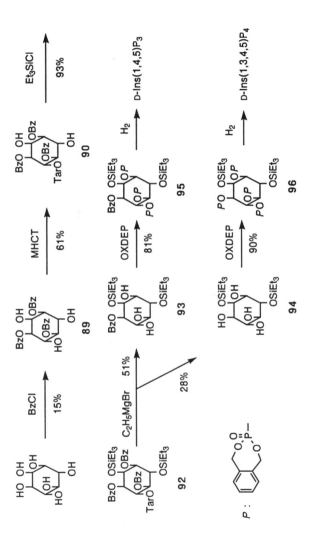

Figure 13. Synthesis of D-Ins(1,4,5)P$_3$ and D-Ins(1,3,4,5)P$_4$ from 1,3,5-tribenzoyl-*myo*-inositol.

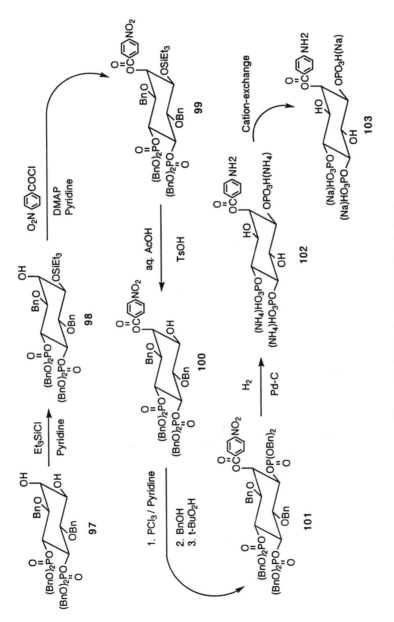

Figure 14. Synthesis of Ins(1,4,5)P$_3$ derivatives.

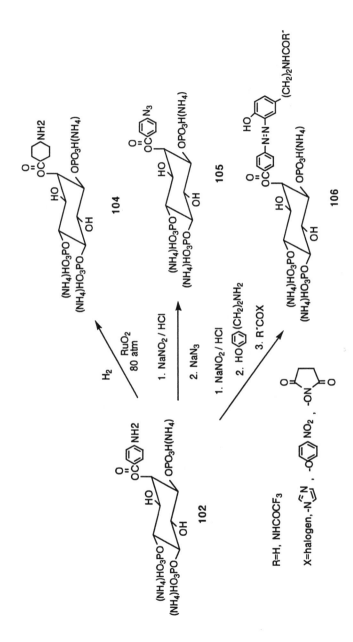

Figure 15. Synthesis of Ins(1,4,5)P₃ derivatives.

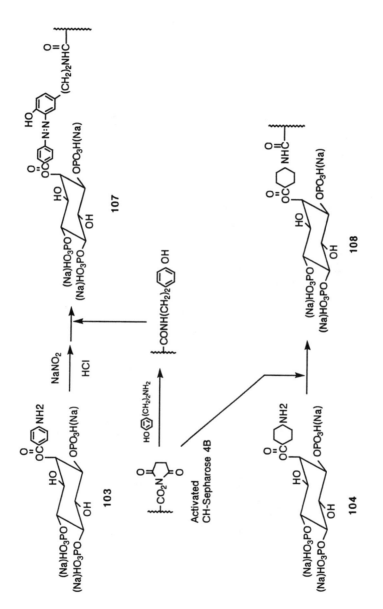

Figure 16. Synthesis of Ins(1,4,5)P$_3$ affinity columns.

azide gave azide **105**. The diazonium salt was also coupled with thyramine and then reacted with acid chloride to give **106**.

The inhibition of Ins(1,4,5)P$_3$-5-phosphatase and Ins(1,4,5)P$_3$-3-kinase and Ca^{2+} release of these compounds (**103, 104, 105,** and **106**) were measured and structure-activity relationships were discussed.[15] Some (**103** and **105**) of them were more potent than the original Ins(1,4,5)P$_3$ in the inhibition of Ins(1,4,5)P$_3$-5-phosphatase.

Activated CH-Sepharose 4B was reacted first with thyramine and then with the diazonium salt of **103** to afford Ins(1,4,5)P$_3$ affinity column **107** (see Figure 16). The compound **104** was coupled directly with activated CH-Sepharose 4B to afford Ins(1,4,5)P$_3$ affinity column **108**. These columns were shown to be very effective tools to isolate Ins(1,4,5)P$_3$ binding protein, Ins(1,4,5)P$_3$-5-phosphatase, and Ins(1,4,5)P$_3$-3-kinase.[16]

Conclusions. Synthetic methods to get *myo*-inositol polyphosphates were developed and improved by combining a new phosphitylating agent (OXDEP), an enantioselective acylating agent (MHCT), and selective acylation techniques. Ins(1,4,5)P$_3$ derivatives and affinity columns were synthesized using the 2-hydroxyl group of *myo*-inositols to afford useful tools to isolate Ins(1,4,5)P$_3$-binding substances.

References

(1) Streb, H.; Irvine, R. F.; Berridge, M. J.; Schulz, I. *Nature* **1983**, *306*, 67. Berridge, M. J.; Irvine, R. F. *ibid.* **1984**, *312*, 315.

(2) Grado, C.; Ballou, C. E. *J. Biol. Chem.* **1961**, *236*, 54. Tomlinson, R. V.; Ballou, C. E. *Ibid.* **1961**, *236*, 1902.

(3) Ozaki, S; Watanabe, Y; Ogasawara, T; Kondo, Y; Shiotani, N; Nishii, H; Matsuki, T. *Tetrahedron Lett.* **1986**, *27*, 3157.

(4) Ozaki, S; Watanabe, Y. *Yukigoseikagakukyokaishi (Journal of Synthetic Organic Chemistry, Jpn)* **1989**, *47*, 363.

(5) Watanabe, Y.; Ogasawara, T.; Shiotani, N.; Ozaki, S. *Tetrahedron Lett.* **1987**, *28*, 2607.

(6) Watanabe, Y.; Nakahira, H.; Bunya, M.; Ozaki, S. *Tetrahedron Lett.* **1987**, *28*, 4179.

(7) Ozaki, S.; Kohno, M.; Nakahira, H.; Bunya, M.; Watanabe, Y. *Chem. Lett.* **1988**, 77.

(8) Ozaki, S.; Kondo, Y.; Nakahira, H.; Yamaoka, S.; Watanabe, Y. *ibid.* **1987**, *28*, 4691.

(9) Watanabe, Y.; Mitani, M.; Ozaki, S. *Chem. Lett.* **1987**, 123.

(10) Watanabe, Y.; Komoda, Y.; Ebisuya, K.; Ozaki, S. *Tetrahedron Lett.* **1990**, *31*, 255.

(11) Watanabe, Y.; Mitani, M.; Morita, T.; Ozaki, S. *J. Chem. Soc., Chem. Commun.* **1989**, 482.

(12) Watanabe, Y.; Shinohara, T.; Fujimoto, T.; Ozaki, S. *Chem. Pharm. Bull.* **1990**, *38*, 562.

(13) Watanabe, Y.; Oka, A.; Shimizu, Y.; Ozaki, S. *Tetrahedron Lett.* **1990**, *31*, 2613.

(14) Watanabe, Y.; Ogasawara, T.; Nakahira, H.; Matsuki, T.; Ozaki, S. *Tetrahedron Lett.* **1988**, *29*, 5259.

(15) Hirata, M.; Yanaga, F.; Koga, T.; Ogasawara, T.; Watanabe, Y.; Ozaki, S. *J. Biol. Chem.* **1990**, *265*, 8404.

(16) Hirata, M.; Watanabe, Y.; Ishimatsu, T.; Yanaga, F.; Koga, T.; Ozaki, S. *Biochem. Biophys. Res. Commun.* **1990**, *168*, 379.

RECEIVED March 11, 1991

Chapter 5

Synthesis of *myo*-Inositol Polyphosphates

Joseph P. Vacca[1], S. Jane de Solms[1], Steven D. Young[1], Joel R. Huff[1],
David C. Billington[2], Raymond Baker[2], Janusz J. Kulagowski[2],
and Ian M. Mawer[2]

[1]Merck Sharp and Dohme Research Laboratories, West Point, PA 19486
[2]Merck Sharp and Dohme Research Laboratories, Neuroscience Research
Centre, Terlings Park, Eastwick Road, Harlow, Essex, CM20 2QR, England

The total synthesis of many naturally occurring myo-inositol phosphates
are reported. The syntheses feature the use of camphanic acid esters for
resolution of protected inositols. The efficient phosphorylation of
hydroxyl groups was achieved by the use of tetrabenzylpyrophosphate
followed by hydrogenolysis of the benzyl protecting groups to obtain the
pure phosphates.

It is well established that phospholipase C stimulation of a number of receptors
results in the hydrolysis of phosphatidylinositol-4,5-bisphosphate [PIP2] giving rise to
diacylglycerol (*1*) and D-*myo*-inositol-1,4,5-trisphosphate [1,4,5-IP3, Fig1] (*2*).
1,4,5-IP3 acts as a second messenger and directly mediates the release of calcium from
intracellular stores (*2*) through activation of specific receptors. The major pathway for
terminating the action of 1,4,5-IP3 is believed to occur through removal of the 5-
phosphate group by a specific 5-phosphatase present in the plasma membranes (*3*) to
afford *myo*-inositol-1,4-bisphosphate (1,4-IP2). Alternatively, phosphorylation of the
3-hydroxyl group of 1,4,5-IP3 via a 3-specific kinase gives 1,3,4,5-IP4 which is then
de-phosphorylated to yield 1,3,4-IP3 (*5*). Eventually, both the 1,4-IP2 and 1,3,4-IP3
are further degraded by other phosphatases giving rise to free inositol via 1- and 4-
monophosphates. The inositol is then recycled to provide more PIP2 (*4-5*).
Inhibition of any events in this pathway could cause interesting effects that may
give insight into new therapeutic targets. For instance, angiotensin II stimulates the
cleavage of PIP2, and the 1,4,5-IP3 released causes an influx of calcium which this
leads to vasoconstriction (*6*). In another case, lithium, a therapy widely used in the
treatment of manic depression, is thought to inhibit the monophosphatase which is
responsible for converting 1-IP and 4-IP into free *myo*-inositol (*7*).
We had an interest in studying the fundamental biochemical processes related to
the inositol pathway and in order to do this, an adequate supply of the natural products
as well as derivatives which were not readily available from natural sources was
needed. At the time our work was initiated, some of the natural products were available
commercially, but only in small amounts and of unknown purity. The supply of D-
myo-2-[3H]-inositol 1,4,5-trisphosphate (obtained by enzymatic conversion of 2-[3H]-
inositol) was expensive and of low specific activity (2-3 Ci/mmol). Our goal was to
develop efficient syntheses of these natural products which could be applicable towards
the synthesis of labelled materials and unnaturally occuring isomers. This review will
outline our successes at achieving these goals.

0097–6156/91/0463–0066$06.00/0
© 1991 American Chemical Society

The key problems encountered in the synthesis of inositol phosphates are (1) synthesis and optical resolution of suitably protected inositol derivatives, (2) efficient phosphorylation of hydroxyl groups, (which is a problem especially when they are vicinal) and (3) deprotection of phosphate substituents in a mild manner in order to facilitate isolation and purification of final products.

Intermediate synthesis. The classical source of protected myo-inositols was in the form of the bis-cyclohexylidene ketals of *myo*-inositol (*8*) (Scheme I) which were obtained via treatment of *myo*- inositol with ethoxy-cyclohexene and a catalytic amount of acid (*9)* The 1,4-diol 4 was crystallized directly from the mixture in 26% yield and the remaining two compounds (**3, 5**) were isolated upon chromatography in 35% and 19% yield respectively. The subsequent manipulation of these compounds leads to a variety of substituted *myo*-inositol derivatives.

A more expeditous route towards *myo*-inositol intermediates is the reaction of *myo*-inositol with triethylorthoformate and a catalytic amount of acid to yield the 1,3,5-protected inositol derivative 6 (*10*). This compound eliminates much of the protecting group sequences needed in manipulating ketals 3 and 4 into suitably protected derivatives.

Resolution of intermediates. In the synthesis of *myo*-inositol 4-phosphate (*11*) (scheme II) it was found that racemic 1,2:4,5-di-O-cyclohexylidene *myo*-inositol (**4**) was selectively alkylated in the 3-position using sodium hydride and benzyl bromide in hot toluene. Compound 7 was treated with S-camphanoyl chloride to yield a mixture of diastereomers **8L** and **8D** which were separable by conventional chromatography. The use of camphanoyl chloride represented a much better techonology for resolving myo-inositol derivatives than the use of mono glycosides (first pioneered by Stepanov) (*12*) and is now utilized in other syntheses of myo-inositol derivatives. Esters **8L** or **8D** were hydrolyzed and then phosphorylated with diphenyl phosphorylchloride (DPPC) to give the 4-monophosphate. Each diastereomer was converted to 4-IP (*6*).

Camphanoyl chloride was also used in our synthesis of D- and L-*myo*-inositol-1,4-bisphosphates **(11b)** (Scheme III). 1,2:4,5-di-O cyclohexylidene-*myo*-inositol (**4**) was treated with two equivalents of (S)-(-)-camphanoyl chloride (DMAP, pyridine) to yield a mixture of diastereomeric bis esters **10D** and **10L** (63%). Attempts to prepare the monoester gave a mixture of mono and bis-esters, and the diastereomeric monoesters were not readily separated by chromatography. The diastereomeric bis-esters were chromatographically separated and then hydrolyzed (LiOH, DME) to give enantiomerically pure diols **4D** and **4L** which were converted to D-(+)-1,4 IP$_2$ and L-(-)-1,4-IP$_2$ as previously outlined by Shvets (*13*).

Myo-inositol trisphosphates. As mentioned earlier, it was anticipated that the hardest problem encountered in the synthesis of *myo*-inositol polyphosphates would be the efficient phosphorylation of compounds containing adjacent hydroxyl groups. Diphenylphosphoryl chloride (DPPC), one reagent widely used to introduce phosphate groups onto isolated hydroxyl groups (*13*), was successfully used by us for synthesizing des-hydroxy 1,4,5-IP3 (*14*) (Scheme IV). Hydroxycyclohexene **(11)** was benzoylated and the olefin functionalized via a Prevost reaction (*15*) with silver benzoate and iodine. Compound 13 was hydrolyzed and reacted with DPPC to afford **14** in good yield. The phenyl groups were then removed via hydrogenation over Adam's catalyst. This was the only case in our hands where DPPC was efficient in phosphorylating vicinal hydroxyl groups. More functionalized substrates (e. g. **5** or **16**) always resulted in poor yields, the main drawback being that the diphenyl phosphate group was unstable during chromatography and subsequent protecting

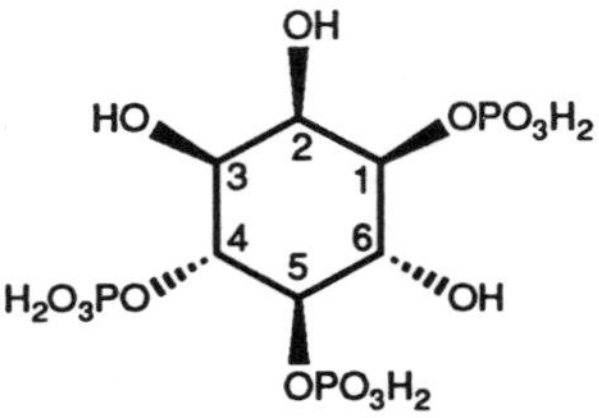

Figure 1. *myo*-Inositol 1,4,5-trisphosphate.

Scheme I

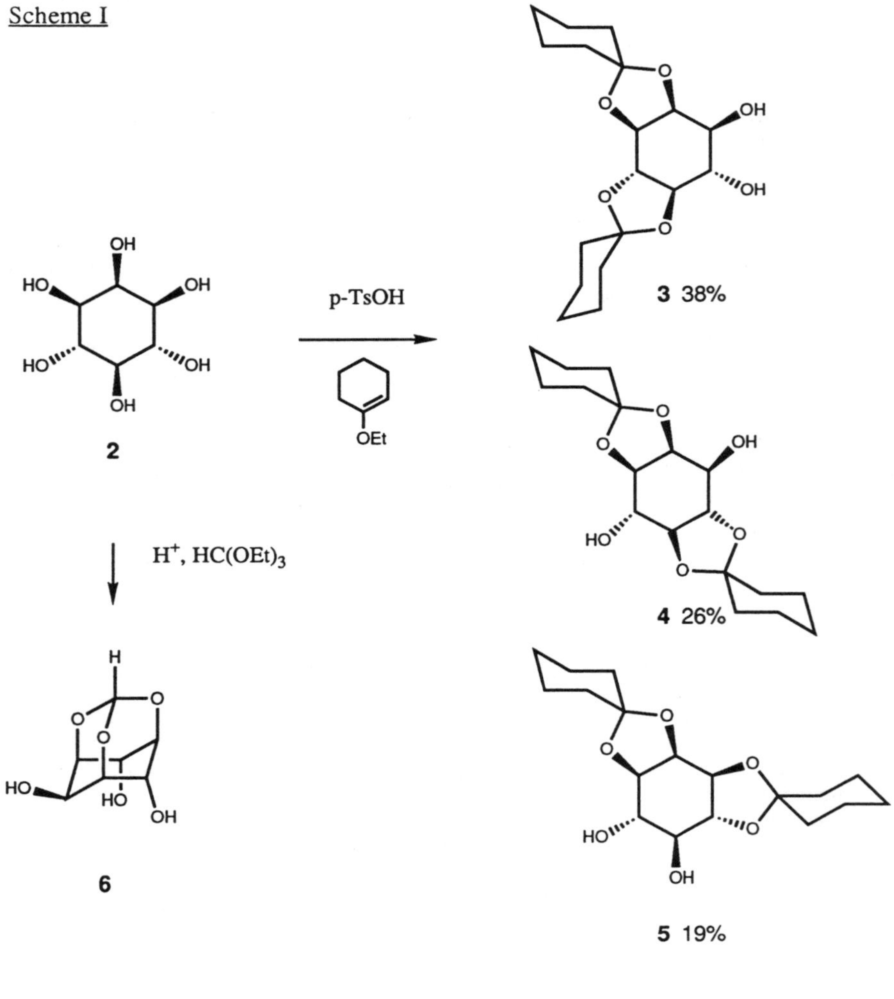

Scheme II

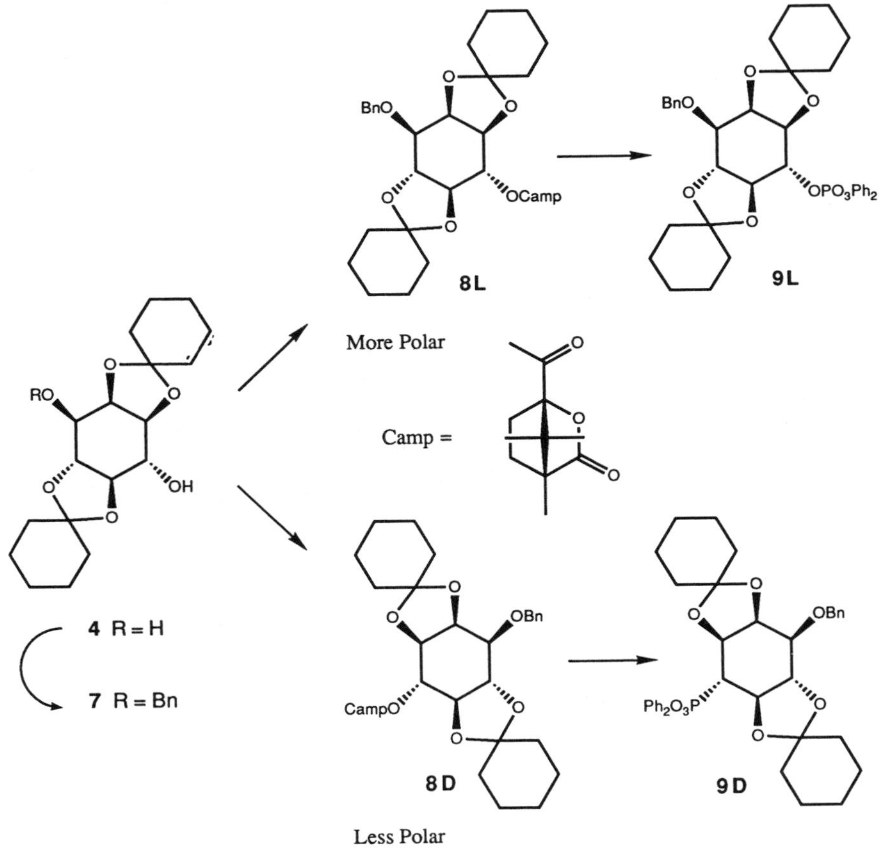
BnO
O
O
OCamp
O
O
8L
OPO3Ph2
9L
More Polar
Camp =
RO
OH
4 R = H
7 R = Bn
OBn
CampO
8D
Ph2O3P
9D
Less Polar

Scheme III

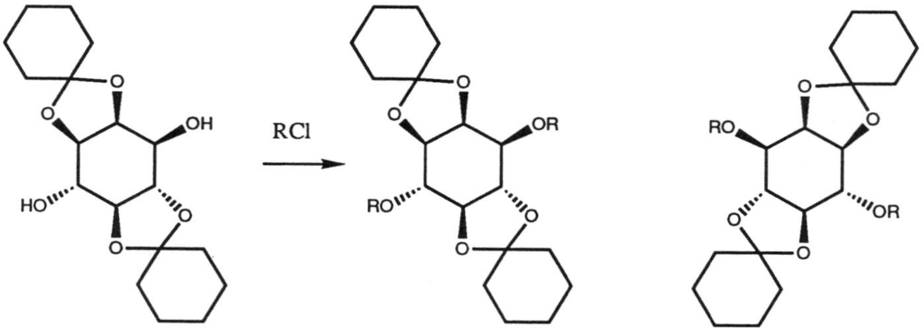

4 (±) **10D** **10L** R = Camp

Scheme IV

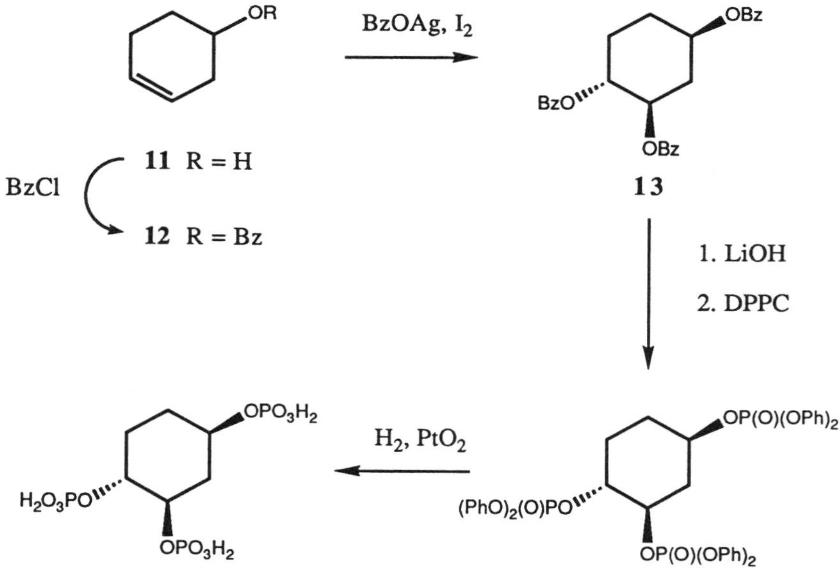

11 R = H

12 R = Bz

BzOAg, I₂

1. LiOH
2. DPPC

H₂, PtO₂

15 **14** **13**

group manipulations. An alternate route for phosphorylating substituted myo-inositol derivatives utilizing a more stable phosphorylating reagent is shown in scheme V.

Diol **16** was reacted with dianilidophosphoryl chloride (*16*) (DAPC) to afford **17** in good yield. The ketal of **17** was removed and the equatorial alcohol successfully phosphorylated with DAPC to afford **19**. All efforts to convert compound **19** to **20** failed due to the harsh deprotection conditions employed. During the course of our investigation. Ozaki et al (*17*) reported on the use of DAPC in the first successful synthesis of 1,4,5-IP3. Their deprotection sequence, however, was reported to proceed in very low yield and the need for an improved procedure was emphasized.

Our solution to the phosphorylation problem is shown in scheme VI. Bartlett (*18*) had reported on the reaction between tetrabenzylpyrophosphate (*19*) (TBPP) and an alkoxide to give a high yield of a dibenzylphosphate. This sequence was first tested in the inositol series with model compound **4**. Compound **4** was deprotonated with n-BuLi and the bis-alkoxide salt was reacted with an excess of TBPP to give a good yield of bisphosphate **21**. However, we were unsure if this method would be applicable to substrates containing vicinally located hydroxyl groups due to the possibility that the anionic alkylation of a vicinal diol would lead to cyclic monophosphates rather than two distinct phosphate groups. To our surprise, a 75% yield of bisphosphate **23** was obtained upon deprotonation of vicinal diol **22** with potassium hydride in refluxing THF which already contained the TBPP. No amount of any cyclic phosphates were observed. With the successful use of TBPP for phosphorylating vicinal diols established, our attention was next focused on synthesizing a substrate for conversion into 1,4,5-IP3.

Garegg (*9b*) (scheme VII) had reported that 4-benzyl-1,2:5,6-di-O cyclohexylidene-*myo*-inositol (**24**) was readily obtained by the phase transfer alkylation of 1,2:5,6-di-O-cyclohexylidene *myo*-inositol (**3**). Following this procedure, a 39% yield of the desired 3-hydroxyl compound **24** was obtained along with a 19% yield of the 4 hydroxyl product **25** and a 9% yield of the previously unreported bis-benzyl compound **26**. All three compounds were easily separated and the undesired ones could be recycled by metal ammonia reduction of the benzyl groups (*20*). Esterification of the free hydroxyl substituent of **24** with (S)-(-)-camphanoyl chloride yielded a mixture of two diastereomers (combined yield = 90%) that were chromatographically separated to give **27D** and **27L**. The diastereomeric purity of each compound exceeded 98% as determined both by HPLC and ^{1}H NMR. All three methyl groups in the ester were singlets, two of which were nonequivalent in the diastereomers. Selective hydrolysis of the trans ketal of each diastereomer gave diols **28D** and **28L**. Basic hydrolysis of the ester group afforded enantiomerically pure triol **29**.

The phosphorylation of compound **29D** was next studied (scheme VIII). Triol **29D** was reacted with tetrabenzylpyrophosphate and potassium hydride in refluxing tetrahydrofuran to give low yields of protected trisphosphate **31D**. Surprisingly, a by-product isolated was epoxide **30D**, which was fully characterized by ^{1}H, ^{31}P NMR, mass spectroscopy and combustion analysis. Presumably, phosphorylation of either the 4- or 5-hydroxyl group was followed by either a second phosphorylation (giving **31D**) or by the adjacent hydroxide attacking the first phosphate in an Sn2 fashion to give epoxide **30D**. This side reaction was suppressed by simply preforming the tri-alkoxide salt at elevated temperatures (THF, 60°C) and then adding the TBPP upon cooling to room temperature. This provided the expected trisphosphate **31D** in 65% yield with no evidence of epoxide or cyclized products. Hydrogenation of trisphosphate **31D** (H2, Pd/C, 95% EtOH) rapidly removed the benzyl groups and mild acid hydrolysis (AcOH, H2O) removed the cyclohexylidene group to give D-1,4,5-myo-inositol trisphosphate in only 8 steps from *myo*-inositol. The main advantage of this sequence was that the protecting groups could be removed using mild

Scheme V

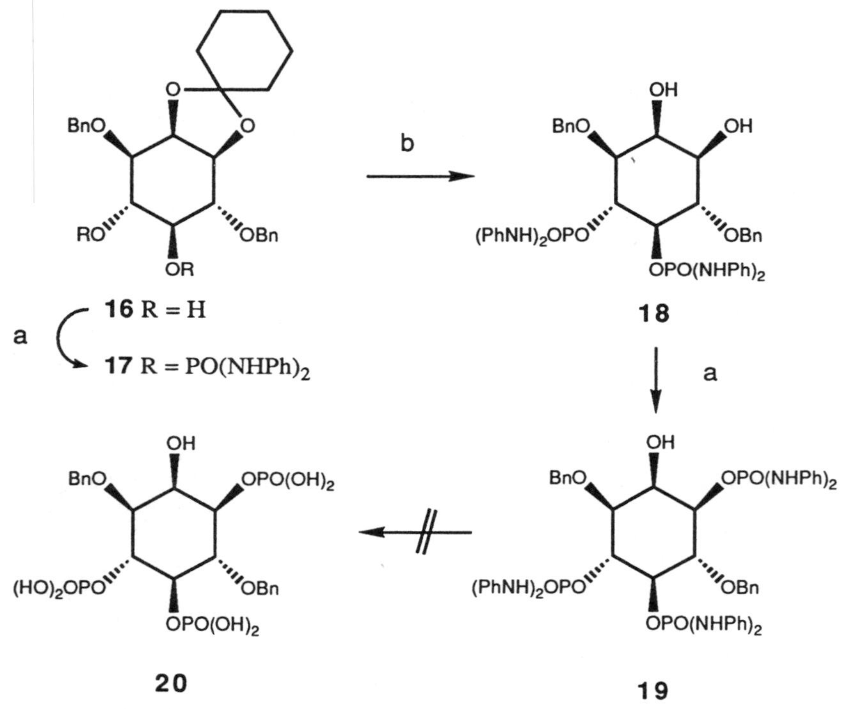

16 R = H

17 R = PO(NHPh)$_2$

18

19

20

Reagents: (a) (PhNH)$_2$POCl, Py; (b) AcOH, H$_2$O, 80°C.

Scheme VI

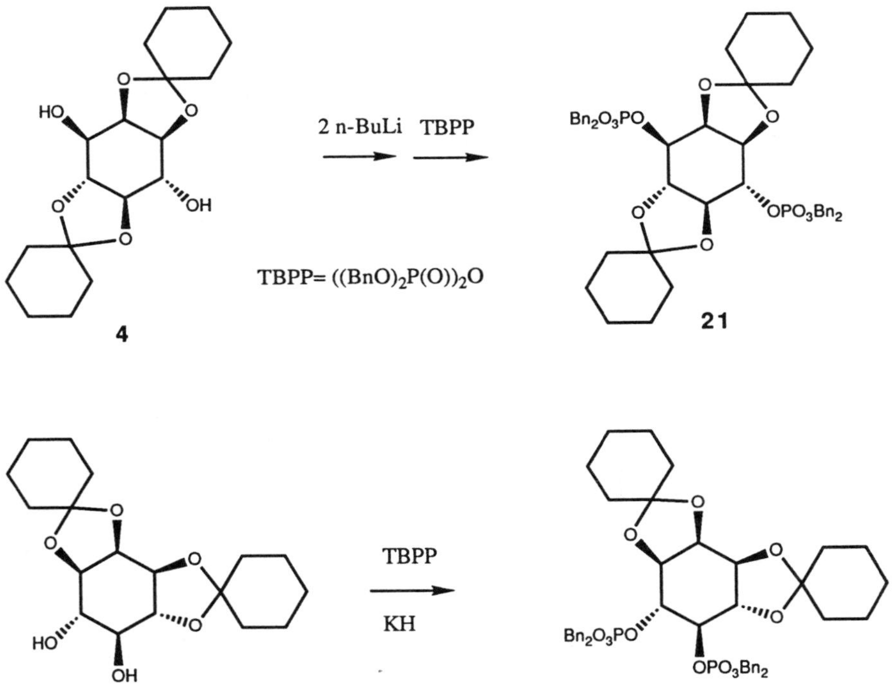

Scheme VII

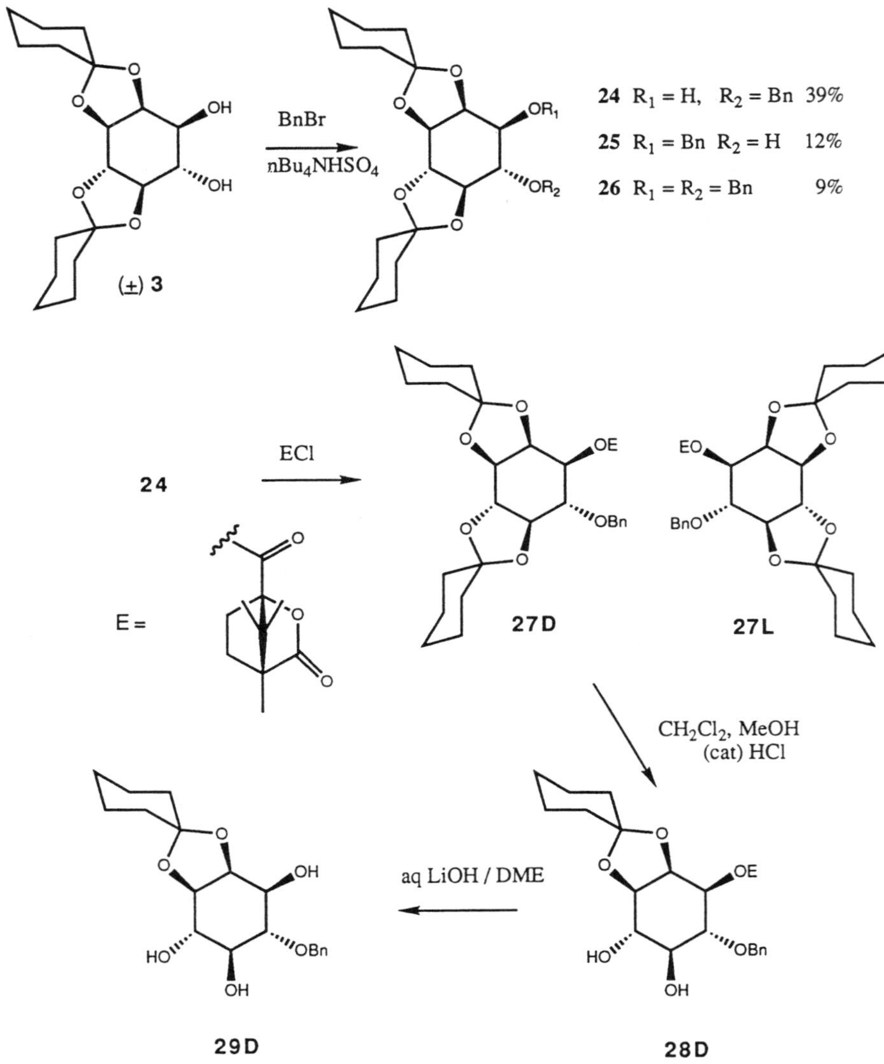
(±) 3
BnBr
nBu4NHSO4
OR1
OR2
24 R1 = H, R2 = Bn 39%
25 R1 = Bn R2 = H 12%
26 R1 = R2 = Bn 9%
24
ECl
E =
OE
OBn
EO
BnO
27D
27L
CH2Cl2, MeOH
(cat) HCl
aq LiOH / DME
OH
HO
29D
28D

Scheme VIII

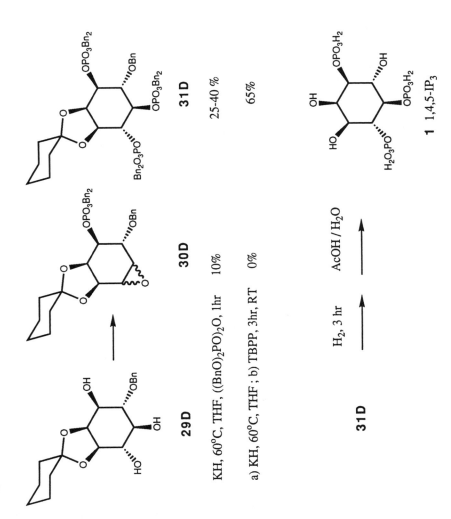

conditions. The 1,4,5-IP$_3$ obtained was essentially pure and did not require any special chromatography to isolate. In a similar manner L-1,4,5-myo-inositol trisphosphate was synthesized starting from camphanate ester **27L**.

Scheme IX shows the synthesis of the unnaturally occurring derivative 6-methoxy 1,4,5-IP$_3$ (**35**). Compound **32** was alkylated in the 4-position with methyl iodode in DMF to afford **33**. The benzyl group was removed usind lithium in ammonia and the alcohol **34** carried on in the usual fashion to afford **35**.

Myo-inositol-2,4,5-trisphosphate (2,4,5-IP$_3$) and *myo*-inositol-1,3,4 trisphosphate (1,3,5-IP$_3$) were synthesized as shown in scheme X. The protected myo-inositol derivatives **35** and **37** were prepared from diol **4** via multi-step procedures previously outlined by Gigg (*11-21*). In contrast to the results obtained in the case of 1,4,5-IP$_3$, treatment of either triol **35** or **37** with an excess of potassium hydride in refluxing THF followed by additon of TBPP at room temperature led to low yields of trisphosphate **36** or **38**. However, when the trisodium salt of **35** (4.5 eq NaH, 1 hr, 0°C) was treated with the pyrophosphate in DMF, a good yield (64%) of **36** was obtained. In the same manner as described for the 1,3,4-IP$_3$ case, compound **37** was phosphorylated and gave tris-phosphate **38** in 70% yield. The nine benzyl groups of either **36** or **38** were removed by catalytic hydrogenation to give (±)-1,3,4-IP$_3$ (75%) and (±) -2,4,5-IP$_3$ (64%). Finally, diol **39** was phosphorylated using the improved conditions to afford **40** in 78% yield.

Myo-**inositol orthoformate.** The bis-acetals **3-5** provided ready access towards various protected *myo*-inositol derivatives and were especially useful when the target molecule posessed free hydroxyl groups in the C-2 and C-3 position that could be protected until the final step using the cyclohexylidene ketal (eg. 1,4,5-IP$_3$ and analogs). However, when the target molecule posessed C-1 and C-3 phosphates, it was always necessary to use extensive protecting methodology to block the C-2 hydroxyl group. To develop a more direct method for protecting the C-2 hydroxyl group early in the synthesis, we explored the use of *myo*-inositol orthoformate (**6**) whose preparation was recently described by Kishi (*10*). This compound provided simultaneous protection of the hydroxyl groups in the C-1, C-3 and C-5 position and was reported to be selectively derivatized in the C-2 position.

Attempts to carry out the selective silylation of the C-2 hydroxyl group of **6** always resulted low yields of the desired product. However, it was discovered that treatment of **6** with one eq. of sodium hydride followed by an eq. of benzyl bromide resulted in preferential alkylation of the C-4 hydroxyl group to give **41** in high yield (22) (scheme XI). Use of a second eq. of each reagent gave compound **42** in which both the C-4 and C-6 axial alcohol were derivatized. One rational for this selectivity is that the initially formed alkoxide salt is coordinated with the adjacent hydroxyl group. Use of other electrophiles led to similar selectivity in 65-80% yields. These intermediates were used to synthesize a number *myo*-inositol phosphates.

Compound **46**, obtained from the reaction of the monoalkoxide of **6** with TBPP was debenzylated and the orthoacetal group hydrolyzed in acid to give *myo*-inositol 4-monophosphate (scheme XII). This route provided 4-IP in only two steps from monoorthoformate **6**.

Myo-inositol 1,3,4,5-tetrakisphosphate was prepared from the mono-alkylated derivatives **44** or **45** (scheme XIII). Protection of the remaining C-2 and C-6 hydroxyl groups as benzyl ethers (**47**) was followed by removal of the C-4 ether and orthoformate group to yield tetraol **48** in good yield. The phosphorylation of **48** was accomplished by treating it with over 4 eq. of sodium hydride followed by excess TBPP. In tetrahydrofuran, either a catalytic amount of imidazole or 18 crown-6 was needed in order to solubilize the tetra-sodio alkoxide salt. No additives were necessary when the reaction was carried out using DMF as the solvent. The ten benzyl groups of

Scheme IX

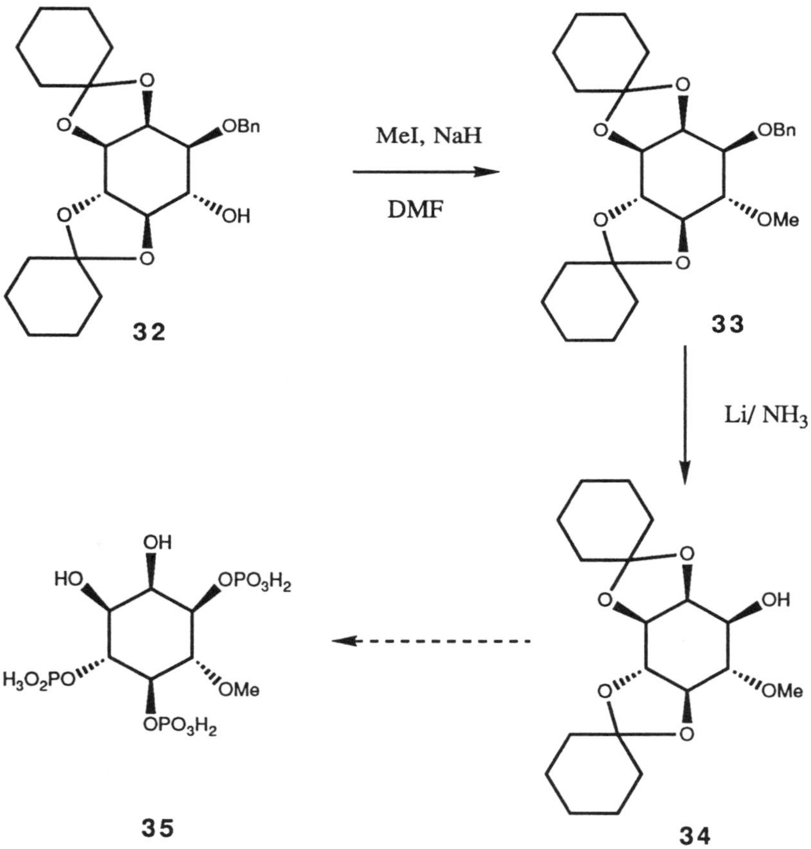

Scheme X. *Reagents and conditions:* (a) NaH, $((BnO)_2P(O))_2O$, DMF, $0°C$;
(b) H_2, 50 psig, 10% Pd/C, 95% EtOH, $25°C$, 3 h.

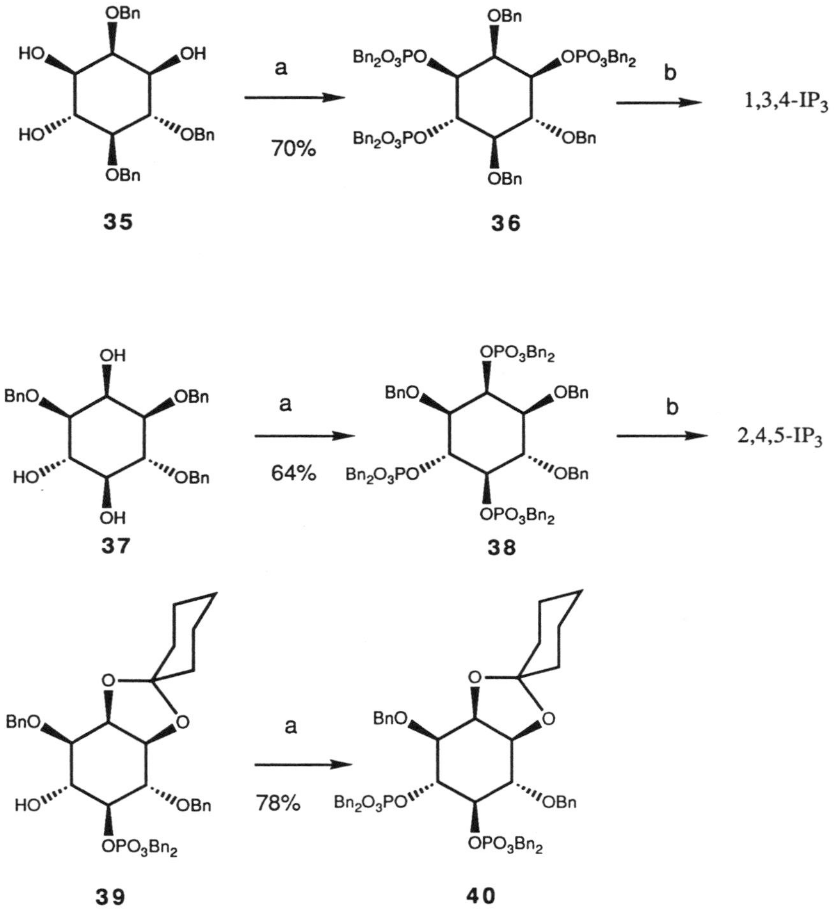

Scheme XI

Elec.	NaH (eq)	R_1	R_2	R_3	CMPD
BnBr	1	Bn	H	H	41
BnBr	2	Bn	Bn	H	42
BnBr	3	Bn	Bn	Bn	43
BnOCH₂Cl	1	BOM	H	H	44
AllylBr	1	Allyl	H	H	45

SCHEME XII

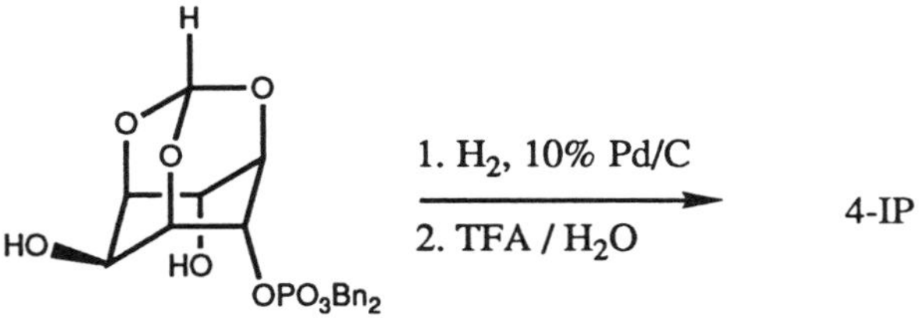

1. H₂, 10% Pd/C

2. TFA / H₂O

4-IP

46

Scheme XIII

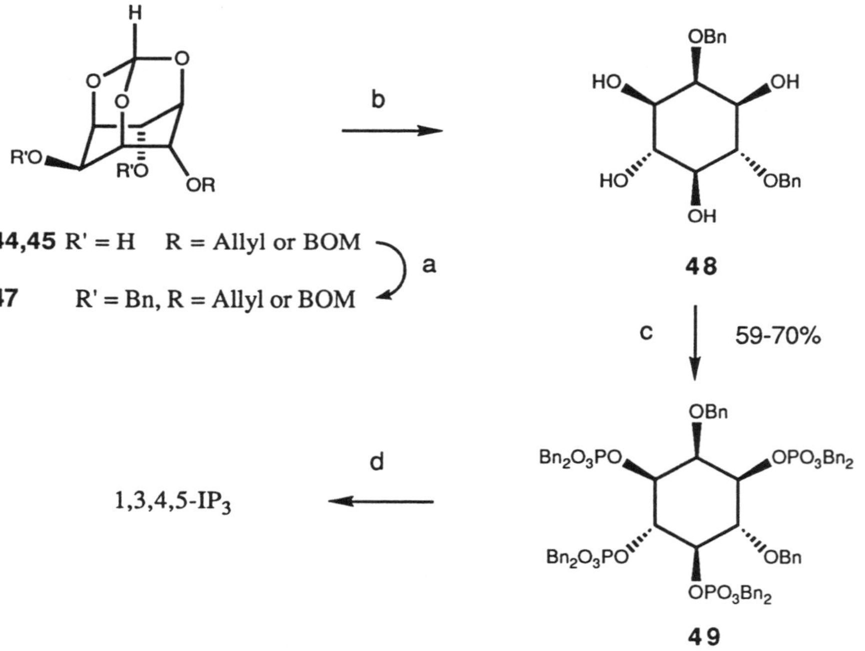
H
O
O
O
R'O
R'O
OR
b
OBn
HO
OH
HO
OBn
OH
44,45 R' = H R = Allyl or BOM
a
47 R' = Bn, R = Allyl or BOM
48
c
59-70%
OBn
Bn2O3PO
OPO3Bn2
d
1,3,4,5-IP3
Bn2O3PO
OBn
OPO3Bn2
49

Reagents: (a) NaH, BnBr, DMF.; (b) 90% EtOH, $RhCl(PPh_3)_3$, DABCO,reflux; 0.1M HCl-MeOH, reflux, 15 min; or TFA, H_2O, 25°C; (c) $((BnO)_2P(O))_2O$,NaH, THF 18-crown-6 or cat imidazole; (d) H_2, Pd/C, 95% EtOH

49 were removed by hydrogenation to afford 1,3,4,5-IP4 in good yield and only 6 steps from *myo*-inositol.

A much shorter synthesis of *myo*-inositol 2-monophosphate (2-IP) which was not contaminated with any of the 1-phosphate was also achieved using compound **6** (scheme XIV). Triol **6** was treated with 2 eq. of sodium hydride followed by 2 eq. of benzyl bromide to give 4,6 bisbenzyl-orthoformate compound **42**. Small amounts of the 2,6 bisbenzyl-ether along with the tribenzyl-ether and 4-monobenzyl-ether were also obtained and could be removed by chromatographic separation. The C-2 alcohol of **42** was phosphorylated (**50**) and the protecting groups removed in the usual fashion to give 2-IP.

Finally, tribenzyl ether **43** provided ready access to 1,3-IP2 and 1,3,5-IP3 (scheme XV). The orthoacetal of **43** was removed with aqueous TFA and the C-1 and C-3 alcohols of **51** were selectively phosphorylated (3/1 ratio of 1,3 vs 1,5) with diphenyl chlorophosphate to afford bisphosphate **53**. Treatment of **53** with lithium in ammonia (-78°C) removed both the phenyl and benzyl protecting groups (*23*) to give the tetralithium salt of 1,3-IP2. The free acid was obtained upon treatment with Amberlite IR-120 and the product isolated as the tetracyclohexylammonium salt.

Exhaustive phosphorylation of triol **51** using TBPP and sodium hydride led to tris phosphate **52** which was converted to 1,3,5-IP3 after hydrogenation of the benzyl groups.

D-*myo*-[1-^{3}H]-inositol   1,4,5-triphosphate.   A critical goal of our program was to develop a synthesis of D-*myo*-^{3}[1-H]-1,4,5-IP3 of high enough specific activity in order to carry out receptor binding studies (*24*). Our route is outlined in scheme XVI.

Enantiomerically pure ester **27D** was hydrolyzed (1M LiOH/DME) providing alcohol **54** in high yield. Swern oxidation of **54** gave ketone **56** which was immediately reduced without further purification. It was found that sodium boryhydride (MeOH, 0°C) gave equatorial alcohol **54** back as the sole product with no detectable amounts of the axial alcohol epimer. Lithium aluminum hydride gave only about a 4:1 ratio of the equatorial/axial alcohol.

To continue the synthesis, compound **54** was first phosphorylated (**57**) and then the less stable trans-ketal was removed (cat. HCl/ MeOH) to afford diol **58**. It was possible to hydrolyze the trans-ketal of **54** to give a triol directly, but we found that in some cases the ketal hydrolysis was hard to control and variable amounts of a pentaol, in which both ketals were removed, was produced.   Diol **58** was bisphosphorylated using TBPP in DMF and the trisphosphate **59** carried on to 1,4,5-IP3 as before.

Following the same route as above but substituting sodium borotritide as the reducing agent, alcohol **55** afforded D-myo-[1-^{3}H]-1,4,5-IP3 having a specific activity of 9.4 Ci/mmol (*25*).

Conclusions.   A number of naturally occurring and unnaturally ocurring *myo*-inositol phosphates were synthesized from two main sources of protected *myo*-inositol derivatives. The efficient resolution of these intermediates was solved by using S-camphanoyl chloride as the resolving agent. Since our original work, many research groups have eliminated the need for resolution of *myo*-inositols by utilizing the chiral pool for protected intermediates and these elegant efforts are outlined in other chapters of this book. The problem of phosphorylation of vicinally substituted hydroxyl groups was solved by using tetrabenzylpyrophosphate with alkoxide anions. The intermediate dibenzylphosphates were stable and simple to purify. This greatly facilitated the protecting group removal sequence and purification of the final products. The strides

Scheme XIV

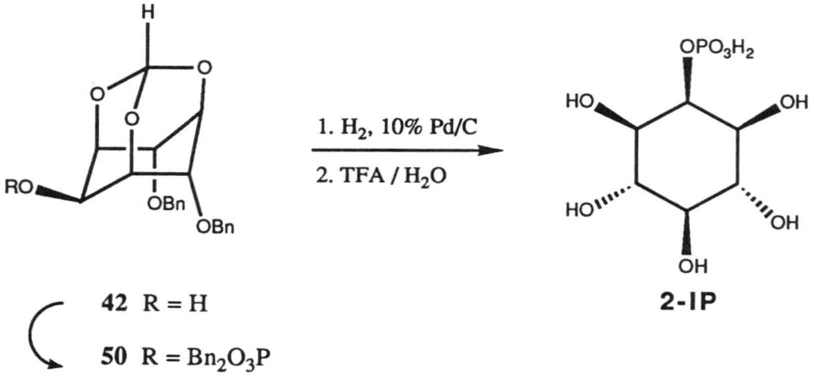

42 R = H

50 R = Bn$_2$O$_3$P

NaH, TBPP

2-IP

Scheme XV

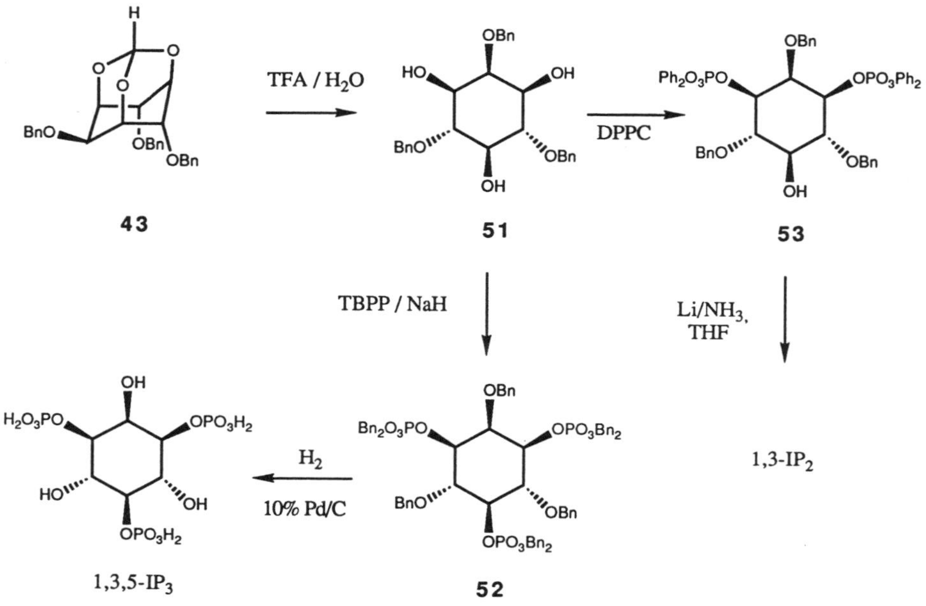

43

51

53

TFA / H$_2$O

DPPC

TBPP / NaH

Li/NH$_3$, THF

1,3-IP$_2$

52

1,3,5-IP$_3$

H$_2$

10% Pd/C

Scheme XVI

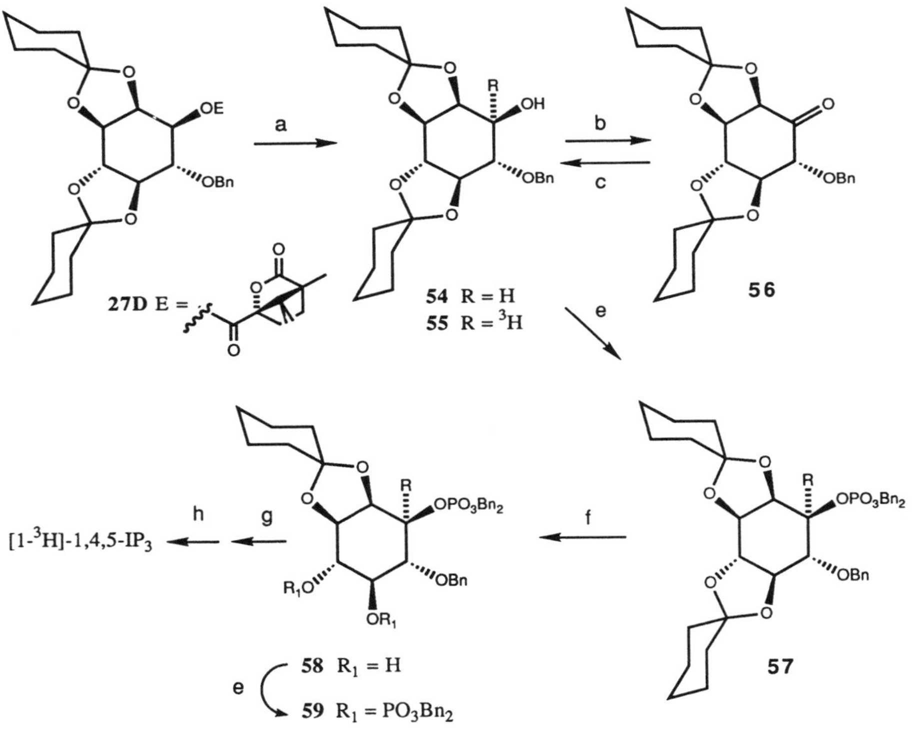

Reagents: (a) LiOH, DME/H$_2$O (1:1), Rt, 4h, (85%); (b) DMSO, COCl$_2$, -78°C, 10 min; NEt$_3$, -78°C, 30 min, (65%); (c) NaBH$_4$, EtOH, RT, 1h, (80%); (d) NaBT$_4$, EtOH, RT, 1h; (e) NaH, DMF, TBPP, 0°C, 3h, (80%); (f) AcCl (cat), MeOH/CH$_2$CL$_2$ (1:9), RT, 3h, (60%); (g) H$_2$, 1 atm EtOH/H$_2$O (9:1), RT, 4h; (h) AcOH/H$_2$O (1:1), RT, 18h, (72%).

made in the synthesis of these compounds hopefully should serve an important role in elucidating the biological role of the *myo*-inositol pathway.

Literature Cited

(1) Nishizuka, Y.; *Nature* **1984**, *3088*, 693.
(2) Berridge, M. J.; Irvine, R. F. *Nature (London)* **1984**, *312*, 315.; Abdel-Latif, A.A. *Pharmacol. Rev.* **1986**, *38*, 227 and references cited therein.
(3) All compounds are drawn as their absolute configuration but are racemic mixtures unless otherwise indicated.
(4) Storey, D. J.; Shears, S. B.; Kirk, C. J.; Michell, R. H. *Nature (London)* **1984**, *312*, 374.
(5) Berridge, M. J. *Biochem. J.* **1984**, *220*, 345.
(6) (a) Griendling, K. K.; Rittenhouse, S. E.; Brock, T. A.; Ekstein, L. S.; Gimbrone, M. A., Jr.; Alexander, R. W. *J. Biol. Chem.* **1986**, *261*, 5901.; (b) Somlyo, A. V.; Bond, M.; Somlyo, A . P.; Scarpa, A. *Proc. Natl. Acad. Sci. U. S. A.* **1985**, *82*, 5231.
(7) Ragan, C. I.; Watling, K. J.; Gee, N. S.; Aspley, S.; Jackson, G. J.; Reid, G. G.; Baker, R.; Billington, D. C.; Barnaby, R. J.; Leeson, P. *Biochem. J.* **1988**, *249*, 314 and references cited.
(8) Angyl, S. J.; Tate, M. E.; Gero, S. D. *J. Chem. Soc. C.*, **1961**, 4116.
(9) (a) Johanissian, A.; Akunian, E. *Bull. Univ. Etat. R. S. S. Armenie* **1930**, *5*, 245.; (b) Garegg, P. J.; Iverson, T.; Johansson, R.; Lindberg, B. *Carbohydr. Res.* **1984**, *130*, 322.
(10) Kishi, Y.; Lee, H. W. *J. Org. Chem.* **1976**, *38*, 3224.
(11) (a) Billington, D. C.; Baker, R.; Kulagowski, J. J.; Mawer, I. M. *J. Chem. Soc., Chem. Comm.* **1987**, 314.; (b) Vacca, J. P.; deSolms, S. J.; Huff, J. R.; Billington, D. C.; Baker, R.; Kulagowski, J. J.; Mawer, I. M. *Tetrahedron* **1989**, *45*, 5679.
(12) Stepanov, A. E.; Klyashchitskii, B. A.; Shvets, V. I.; Evstigneeva, R. P. *Biorg. Khim.* **1976**, *2*, 1627.
(13) Krylova, V. N.; Kobel'kova, N. I.; Olenik, G. F.; Shvets, V. I. *J. Org. Chem. USSR* **1980**, *16*, 59.
(14) Polokoff, M. A.; Bencen, G. H.; Vacca, J. P.; deSolms, S. J.; Young, S. D.; Huff, J. R. *J. Biol. Chem.* **1988**, *263*, 11922.
(15) McCasland, G. E.; Horswill, E. C. *J. Am. Chem. Soc.* **1954**, *76*, 1654.
(16) (a) Cremlyn, R. J. W.; Dewhurst, B. B.; Wakeford, D. H. *J. Chem. Soc. C* **1971**, 300.; (b) Krylova, V. N.; Gornaeva, N. P.; Olenik, G. F.; Shvets, V. I. *J. Org. Chem. USSR* **1980**, *16*, 277.; (c) Krylova, V. N.; Lyutik, A. I.; Gornaeva, N. P.; Shvets, V. I. *J. Gen. Chem. USSR* **1981**, *51*, 183.
(17) Ozaki, S.; Watanabe, Y.; Ogasawara, T.; Kondo, Y.; Shiotani, N.; Nishii, H.; Matsuki, T. *Tetrahedron Lett.* **1986**, *26*, 3156.
(18) Chouinard, P. M.; Bartlett, P. A. *J. Org. Chem.* **1986**, *51*, 75.
(19) Khorana, H. G.; Todd, A. R. *J. Chem. Soc.* **1953**, 2257.
(20) Fraser-Reid has recently reported a chemoselective alkylation of diol 3 to give exclusively the 3-hydroxyl compound 24. (a) Yu, K.; Fraser-Reid, B. *Tetrahedron Lett.* **1988**, *26*, 979.; (b) Yu, K.; Ko, K.; Fraser-Reid, B. *Synthetic Comm.* **1988**, *18*, 465.
(21) Gigg, J.; Gigg, R.; Payne, S.; Conant, R. *Carbohyd. Res.* **1985**, *140*, C1-C3; (b) Gigg, J.; Gigg, R.; Payne, S.; Conant, R. *J. Chem. Soc. Perkin Trans. I* **1987**, 423.
(22) Billington, D. C.; Baker, R.; Kulagowski, J. J.; Mawer, I. M.; Vacca, J. P.; deSolms, S. J.; Huff, J. R. *J. Chem. Soc. Perkin Trans I* **1989**, 1423.

(23) Maryanoff, B. E.; Reitz, A. B.; Tutwiler, G. F.; Benkovic, S. J.; Benkovic, P. A.; Pilkis, S. J. *J. Am. Chem. Soc.* **1984**, *106*, 7851.

(24) For a recent synthesis of D-myo-[1-^{3}H]-1,4,5-IP$_3$ see Maracek, J. F.; Prestwich, G. D. *J. Labelled Comp.* **1989**, *27*, 917.

(25) Polokoff, M. unpublished results. This compound is now commercially available (NEN Products) and is prepared in higher specific activity form, using the above procedure.

RECEIVED March 11, 1991

Chapter 6

Preparation of Optically Active *myo*-Inositol Derivatives as Intermediates for the Synthesis of Inositol Phosphates

Trupti Desai[1], Alfonso Fernandez-Mayoralas[2], Jill Gigg[1], Roy Gigg[1], Carlos Jaramillo[2], Sheila Payne[1], Soledad Penades[2], and Nathalie Schnetz[3]

[1]**Laboratory of Lipid and General Chemistry, National Institute for Medical Research, Mill Hill, London NW7 1AA, England**
[2]**Instituto de Quimica Organica General, C.S.I.C., Madrid, Spain**
[3]**Laboratoire de Chimie Organique, Faculté de Pharmacie, Université Louis Pasteur, Strasbourg, France**

The following racemic *myo*-inositol derivatives: 1,3-di-O-allyl-2,6--di-O-benzyl-; 1,3,4-tri-O-allyl-2-O-benzyl-; 1,5-di-O-benzyl-2,3-O--isopropylidene-4-O-*p*-methoxybenzyl-; 2,6-di-O-benzyl-1,5-di-O-*p*--methoxybenzyl-; 2,4-di-O-benzyl-5,6-O-isopropylidene-1-O-*p*-methoxybenzyl-; 1,2,4-tri-O-benzyl-5,6-O-isopropylidene-; 1,5,6-tri--O-benzyl-2,3-O-isopropylidene-; 2,5,6-tri-O-benzyl-1-O-*p*-methoxybenzyl- and 1,2,5,6-tetra-O-benzyl- were found to be readily resolved by crystallisation of the (+)- or (-)- ω-camphanates. The chiral inositol derivatives prepared from these diastereoisomeric camphanates are suitable intermediates for the synthesis of most of the inositol phosphates and phosphatidylinositol phosphates of the 'phosphatidyl--inositol cycle' as well as the 'lipid anchor' of cell surface glycoproteins, the serologically active glycolipids of *Mycobacterium tuberculosis* and the plant glycolipid 'phytoglycolipid'.

This review will deal mainly with unpublished work and will refer only briefly to recently published papers (*1-6*) from Mill Hill concerned with inositol chemistry. Some of the work described forms part of an International Patent Application (*7*) and of the M.Phil. thesis of Trupti Desai (*8*).

In the formulae, racemic inositol derivatives are indicated by (±) in the ring and chiral inositol derivatives, represented in their correct absolute configurations are shown with thickened lines in the ring.
All = -CH$_2$-CH=CH$_2$; Crot = -CH$_2$-CH=CH-Me; Prop = -CH=CH-Me;
Bn = -CH$_2$Ph; pMB = -CH$_2$Ph(*p*OMe);
(-)-camph and (+)-camph = (-)- and (+)-ω-camphanate esters;

Ⓟ = PO(OH)$_2$; |P| = PO(OBn)$_2$;   △P = PO(OCH$_2$CH$_2$CN)$_2$.

Dedicated to Professor Herbert E. Carter on the occasion of his 80th birthday (September 25th 1990).

Most of the technology required for the synthesis of *myo*-inositol phosphates (protection of specific hydroxyl groups, phosphorylation and deprotection methods) has been elaborated by several groups active in this area (*9,10*). The major problem remaining has been the efficient preparation of intermediates for the synthesis of the chiral inositol phosphates. By the investigation of many derivatives of *myo*-inositol, we have developed resolutions [using commercially available (+)- and (-)-ω-camphanic acids] which give the required diastereoisomers in good yield by crystallisation and, with these chiral intermediates available, most of the natural inositol phosphates and phosphatidylinositol phosphates, so far described, can be prepared.

Intermediates for the Synthesis of Inositol Phosphates.

We have described (*4,5*) the synthesis of racemic 1,2,4-tri-O-benzyl-5,6-O-isopropylidene-*myo*-inositol (**3**) *via* the allyl ether (**1**) and this synthesis has now been improved by using the crotyl ether [**2**, m.p. 109°, prepared in a similar way to (**1**)] as an intermediate. The removal of the crotyl group (*11*) is a one stage reaction whereas removal of the allyl group requires two stages.

Our first attempt (*4,7*) to resolve an inositol derivative (**3**) as the (-)-ω--camphanate was successful giving a high yield of one diastereoisomer [**4**, m.p.172°, [α]$_D$ +53°(CHCl$_3$)] by crystallisation whilst the other (**5**) was obtained as a syrup. Unfortunately the crystalline diastereoisomer (**4**) was not the one required for the preparation of 2,3,6-tri-O-benzyl-D-*myo*-inositol (**10**) (to be used as an intermediate for the synthesis of D-*myo*-inositol 1,4,5-trisphosphate) but on hydrolysis gave the enantiomer as shown by the conversion of (**4**) into (+)-bornesitol (**8**). However the crude syrupy diastereisomer (**5**) was converted into the crude allyl ether (**7**) which after deacetonation and crystallisation gave the enantiomerically pure diol [**9**, m.p. 98°, [α]$_D$ +20°(CHCl$_3$)] because this crystallises much more readily than the racemate. Its purity was established by comparison of the rotation with that of its enantiomer (**6**) and by 200MHz ^{1}H-n.m.r. spectroscopy of the diacetates in the presence of the chiral shift reagent [Eu(hfc)$_3$] at different concentrations. The chiral allyl ether (**9**) is a useful intermediate for the synthesis of the 5-phosphorothioate analogue of D-*myo*-inositol 1,4,5-trisphosphate (see below). We also showed that conversion of the crude (-)-ω-camphanate (**5**) into the crude bis-(-)-ω-camphanate (**11**) and crystallisation gave the pure diastereoisomer [**11**, m.p. 112°, [α]$_D$ +13° (CHCl$_3$). The diol derived by saponification of (**11**) had m.p. 106°, [α]$_D$ +15° (CHCl$_3$)].

In 1989, (+)-ω-camphanic acid became commercially available (Fluka) and we were therefore able to convert the alcohol obtained by saponification of the crude syrupy (-)-ω-camphanate (**5**) into the beautifully crystalline (+)-ω-camphanate (**13**) which is the enantiomer of the crystalline (-)-ω-camphanate (**4**). This approach was used rather than treating the alcohol (**3**) directly with (+)-ω-camphanic acid chloride in order to use less of the more expensive (+)-ω-camphanic acid [seven times the price of the (-)-isomer (Fluka)] and also because the diastereoisomer (**4**) was required for the synthesis of other chiral inositol phosphates (see below). Thus three routes were available, from the racemic alcohol (**3**), for the preparation of 2,3,6-tri--O-benzyl-D-*myo*-inositol [**10**, m.p. 120°, [α]$_D$ +10°(CHCl$_3$)] to be used as an intermediate for the synthesis of D-*myo*-inositol 1,4,5-trisphosphate.

We have subsequently used the camphanic acids in large quantities for resolutions and as they are expensive we were interested in their recovery. A standard text on the subject of resolutions (*12*, see also *13*) stated that for camphanates: "Regeneration of the resolved alcohols is affected by saponification or by reduction with lithium aluminium hydride; in either process the camphanic acid is destroyed. This fact constitutes a significant limitation to the use of this resolving

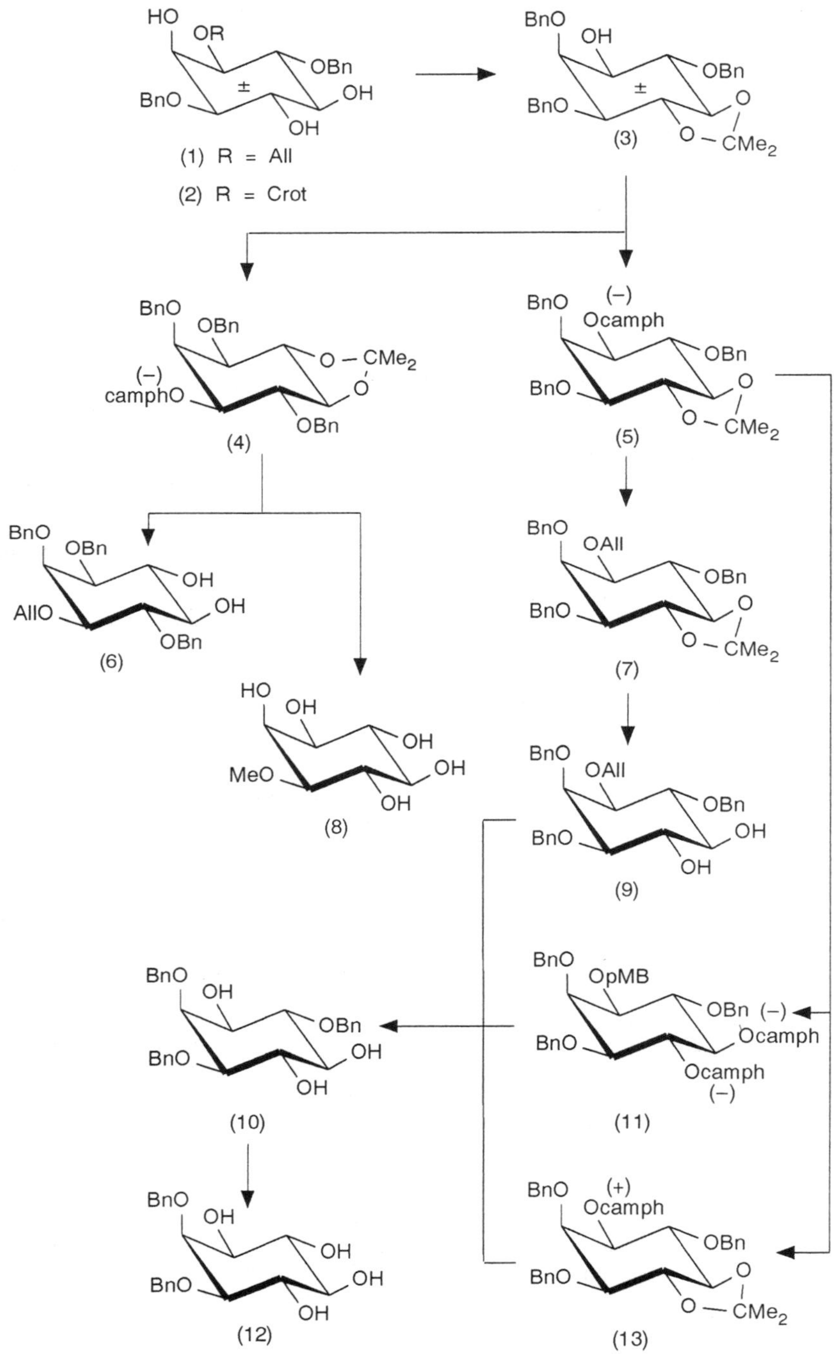
HO
OR
OBn
OH
BnO
±
OH
(1) R = All
(2) R = Crot
BnO
OH
OBn
O
BnO
±
O – CMe2
(3)
BnO
OBn
O – CMe2
(–)
camphO
O
OBn
(4)
(–)
BnO
Ocamph
OBn
O
BnO
O – CMe2
(5)
BnO
OBn
OH
AllO
OH
OBn
(6)
BnO
OAll
OBn
O
BnO
O – CMe2
(7)
HO
OH
OH
MeO
OH
OH
(8)
BnO
OAll
OBn
BnO
OH
OH
(9)
BnO
OH
OBn
BnO
OH
OH
(10)
BnO
OpMB
OBn (–)
Ocamph
BnO
Ocamph
(–)
(11)
BnO
OH
OH
BnO
OH
OH
(12)
BnO
(+)
Ocamph
OBn
O
BnO
O — CMe2
(13)

agent on any but a small scale." However, we find that acidification of the aqueous layer (containing the hydrolysed camphanic acid) after the saponification of the camphanate, and subsequent extraction with dichloromethane, allows the recovery of camphanic acid in high yield. This recovery procedure has also been described by Vogel (14,15). In the separation of the diastereoisomeric camphanates by crystallisation an inspection of the methyl resonances of the camphanate groups in the ^{1}H-n.m.r. spectrum (even at 90MHz) is valuable since the chemical shifts are usually sufficiently different for each diastereoisomer to allow an assessment of purity.

The diastereoisomers (13 and 4) are also suitable intermediates for the synthesis of D-*myo*-inositol 1- and 3-phosphates respectively. Also after removal of the camphanate groups from (13) and (4) and subsequent benzylation and deacetonation they provide 1,2,3,6- and 1,2,3,4-tetra-O-benzyl-D-*myo*-inositols respectively, suitable for the synthesis of D-*myo*-inositol 4,5- and 5,6-bisphosphates. 1,2,3,4-Tetra-O-benzyl-D-*myo*-inositol has m.p. 106°, $[\alpha]_D$ -15°(CHCl$_3$) and the derived acetonide has m.p. 84°,$[\alpha]_D$ +31°(CHCl$_3$). We have also shown that the 6-O-benzyl ether of 2,3,6-tri-O-benzyl-D-*myo*-inositol (10) can be preferentially removed by acetolysis, following the method developed by Angyal (16), to give 2,3-di-O-benzyl-D-*myo*-inositol (12) which is a suitable intermediate for the synthesis of D-*myo*-inositol 1,4,5,6-tetrakisphosphate. 1,2-Di-O-benzyl-D-*myo*-inositol can be prepared in the same way from (4) for use as an intermediate for D-*myo*-inositol 3,4,5,6-tetrakisphosphate. Both of these tetrakisphosphates are of current interest (17-20) as intermediates in the metabolism of inositol phosphates.

Because of the success achieved with the resolution of the alcohol (3) *via* the camphanates, we decided to investigate the resolution of analogues of (3) in which one of the benzyl groups was replaced by a *p*-methoxybenzyl group. In the case of the racemic 2,6-di-O-benzyl-3-O-*p*-methoxybenzyl derivative [14, prepared from racemic 1,2:4,5-di-O-isopropylidene-3-O-*p*-methoxybenzyl-*myo*-inositol (6) in a similar way to the preparation of (3)] identical crystallisation behaviour to that of (3) was observed giving easy access to the two crystalline diastereoisomers [15, m.p. 152°, $[\alpha]_D$ -52°(CHCl$_3$)] and its enantiomer (16). With the other two mono-*p*-methoxybenzyl ether analogues of (3) no good separation of the diastereoisomers by crystallisation has yet been achieved. After removal of the camphanate and isopropylidene groups from the chiral *p*-methoxybenzyl ether (15) the product [17, R = H, m.p. 122°, $[\alpha]_D$ +10°(CHCl$_3$)] is a suitable intermediate for the synthesis of D-*myo*-inositol 1,4,5-trisphosphate. Removal of the *p*-methoxybenzyl group from (17, R = H) gave 2,6-di-O-benzyl-D-*myo*-inositol [21, m.p. 148°,$[\alpha]_D$ -28°(CHCl$_3$), $[\alpha]_D$ -28°(EtOH)] which has been prepared previously (21-24). This compound has also been prepared by other routes (see below) and is a suitable intermediate for the synthesis of D-*myo*-inositol 1,3,4,5-tetrakisphosphate.

Allylation of the alcohol derived from (15) by saponification and subsequent removal of the *p*-methoxybenzyl and isopropylidene groups gave the triol [18, m.p. 110°, $[\alpha]_D$ +4°(CHCl$_3$)]. Compound (15) was also converted into the diol (19) and partial de-*p*-methoxybenzylation of this gave the triol [22, m.p. 137°,$[\alpha]_D$ -9° (CHCl$_3$)]. Both compounds (18 and 22) are required as intermediates for the synthesis of phosphatidylinositol 3,4,5-trisphosphate (see below). Compound (17) is also being investigated as an intermediate for the synthesis of the 3-phosphorothioate analogue of D-*myo*-inositol 1,3,4,5-tetrakisphosphate *via* the alcohol (20). This phosphorothioate should not be metabolised to D-*myo*-inositol 1,4,5-trisphosphate, as is D-*myo*-inositol 1,3,4,5-tetrakisphosphate itself (25), and should therefore be useful in investigating the function of the latter as a 'second messenger' (26,27). Compound (20) is also being investigated as an intermediate for the preparation of ^{3}H-labelled D-*myo*-inositol 1,4,5-trisphosphate (see below).

In order to make some of the intermediates described above more readily available we have developed new routes involving the one-pot tin-mediated allylation of *myo*-inositol (23) in the presence of tetrabutylammonium bromide with acetonitrile as a solvent and molecular sieve 3A in a Soxhlet to remove the water liberated. With an excess of reagents the *myo*-inositol was converted (during 24 hours) into a readily separable mixture of the two tetra-O-allyl inositols (26 and 27) together with small amounts of penta-O-allyl inositols and the symmetrical 1,3,5--tri-O-allyl-*myo*-inositol (m.p. 69°, triacetate m.p. 125°). The tetra-O-allyl inositols (26 and 27) were obtained as syrups which gave crystalline diacetates (m.p. 90° and 75° respectively) and on partial acetylation at 20° gave the mono-acetates with a free axial 2-hydroxyl group (m.p. 70° and 102° respectively) in good yield. These mono-acetates are potentially useful compounds for synthetic work. We had previously (7) prepared the mixture of tetra-O-allyl inositols (26 and 27) by applying the tin-mediated allylation to a mixture of 1,4-, 4,5- and 1,6-di-O-allyl-*myo*--inositols. When *myo*-inositol was subjected to the tin-mediated allylation using three equivalents of dibutyltin oxide the racemic 1,3,4-tri-O-allyl-*myo*-inositol (24, m.p. 85°) was isolated as a major product and this also gave a crystalline tri-acetate (m.p. 137°).

Allylation of *myo*-inositol under these conditions with 2.2 equivalents of dibutyltin oxide gave a complex mixture of water soluble mono-, di- and tri-O-allyl--*myo*-inositols which were most readily separated and characterised after their conversion into ether soluble isopropylidene derivatives. Column chromatography of these gave racemic 1-O-allyl-2,3;4,5-di-O-isopropylidene- (m.p.119°, acetate m.p. 97°), 1-O-allyl-2,3:5,6-di-O-isopropylidene- (3), 3,4-di-O-allyl-1,2;5,6-di-O--isopropylidene- (5) and 1,3,4-tri-O-allyl-5,6-O-isopropylidene-*myo*-inositols all in *ca.* 15% yield. The major product (*ca.* 25% yield) was 1,3-di-O-allyl-4,5-O--isopropylidene-*myo*-inositol which gave 1,3-di-O-allyl-*myo*-inositol (m.p. 121°) on hydrolysis. We had hoped this would be formed in higher yield as it is a useful compound particularily for the synthesis of racemic 2,4-di-O-benzyl-*myo*-inositol (28). Minor products were identified as 1,4-di-O-allyl-2,3;5,6-di-O-isopropylidene-(8%,3), 1,3,5-tri-O-allyl- (5%) and 1,5-di-O-allyl-2,3-O-isopropylidene-*myo*-inositol (2%, m.p. 92°). In the above tin-mediated reactions allyl bromide was substituted by crotyl bromide to give similar results.

The 1,3,4-tri-O-allyl-5,6-O-isopropylidene-*myo*-inositol (33) described above was also prepared from the 1,3,4-tri-O-allyl-*myo*-inositol (24) obtained in the tin mediated allylation of *myo*-inositol using three equivalents of dibutyltin oxide. Benzylation of compound (33) and subsequent deacetonation gave racemic 1,3,4-tri--O-allyl-2-O-benzyl-*myo*-inositol (34, m.p. 55°, diacetate m.p. 123°). The bis-ω--camphanates of the diol (34) were separated by crystallisation and from the (-)-ω-camphanate [m.p. 211°, $[\alpha]_D$ +7°(CHCl$_3$)] the chiral diol [37, acetate m.p. 135°, $[\alpha]_D$ +24°(CHCl$_3$)] was obtained and its enantiomer (38) was obtained preferentially by crystallisation of the bis-(+)-ω-camphanate of (34). Benzylation of compound (38) and deallylation gave 2,5,6-tri-O-benzyl-D-*myo*-inositol [35, m.p. 105°, $[\alpha]_D$ -27°(CHCl$_3$)] suitable as an intermediate for the synthesis of D-*myo*--inositol 1,3,4-trisphosphate. *p*-Methoxybenzylation of (37 and 38) followed by deallylation, acetonation, benzylation, deacetonation and de-*p*-methoxybenzylation gave 2,3-di-O-benzyl-D-*myo*-inositol (36) and its enantiomer (41) respectively. This is an alternative route to that described above based on the acetolysis of compound (10) and its enantiomer.

Benzylation of the dibutylstannylene derivative of (38) gave (together with the regioisomer) 1,3,4-tri-O-allyl-2,6-di-O-benzyl-D-*myo*-inositol [40, acetate m.p. 82°, $[\alpha]_D$ +22°(CHCl$_3$)] which on *p*-methoxybenzylation and deallylation gave the triol [39, R=H, m.p. 115°, $[\alpha]_D$ -22°(CHCl$_3$)] which we intend to use as an intermediate for the synthesis of the 5-phosphorothioate analogue of D-*myo*-inositol

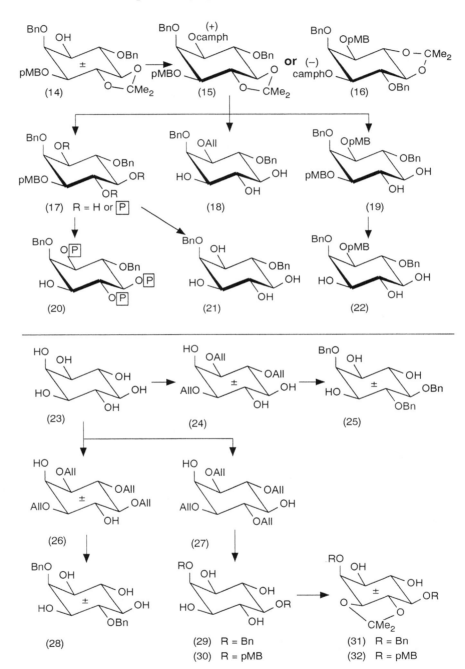

1,3,4,5-tetrakisphosphate *via* the protected phosphate [**39**, R =⟨P⟩, syrup, $[\alpha]_D$ 0°
(CHCl$_3$)]. This 5-phosphorothioate analogue should be stable to the 5-phosphatase
that normally metabolises D-*myo*-inositol 1,3,4,5-tetrakisphosphate to the 1,3,4-
-trisphosphate and could be a useful compound for investigating the proposed role
(*26,27*) of the 1,3,4,5-tetrakisphosphate as a 'second messenger'. The racemic
1,3,4,5-tetra-O-allyl-*myo*-inositol (**26**) gave a mixture of diastereoisomeric
(-)-ω-camphanates which could be separated by chromatography and the slow (t.l.c.)
(-)-ω-camphanate [m.p. 169°, $[\alpha]_D$ -1°(CHCl$_3$)] also crystallised preferentially from
the mixture. This on saponification and subsequent methylation and deallylation
gave 2,4-di-O-methyl-D-*myo*-inositol identical with the material prepared from
1,3,5,6-tetra-O-benzyl-D-*myo*-inositol (*6*) thus establishing the absolute
configuration of this diastereoisomer.

   Benzylation of the tetra-O-allyl derivative (**26**) and subsequent deallylation
using Pd/C (*28*) gave the racemic 2,4-di-O-benzyl-*myo*-inositol (**28**, m.p. 126°). We
get variable yields in this reaction due to debenzylation and are investigating other
procedures for deallylation. Treatment of the di-O-benzyl ether (**28**) with acidic
dimethoxypropane gave a mixture of the isopropylidene derivatives (**42**, m.p. 96°,
acetate m.p. 118°) and (**43**, m.p. 110°) which were readily separated by silica gel
chromatography. Compound (**43**) is a suitable intermediate for the synthesis of
myo-inositol 1,3-bisphosphate. Partial *p*-methoxybenzylation of (**43**) gave the useful
intermediate (**14**). Allylation of (**43**) and subsequent deacetonation gave the diol
(**45**, isopropylidene derivative m.p. 106°; diacetate m.p. 129°). Crystallisation of the
bis-(-)-ω-camphanates of the diol (**45**) gave a pure diastereoisomer (m.p. 163°)
which gave the chiral diol (**44**) on saponification. Deallylation of (**44**) gave 2,6-di-
-O-benzyl-D-*myo*-inositol (**21**) identical with the material described above. We
showed previously (*6*) that crystallisation of the bis-(-)-ω-camphanates of the
racemic 1,2,5,6-tetra-O-benzyl-*myo*-inositol (**46**) gave the pure bis-camphanate of
the chiral alcohol (**47**) and this diol is a suitable intermediate for the synthesis of
D-*myo*-inositol 3,4-bisphosphate. We therefore decided to investigate the behaviour
of analogues of the tetra-O-benzyl ether (**46**) containing some *p*-methoxybenzyl
ether groups in place of benzyl groups to see if the (-)-ω-camphanates of these
would also separate by crystallisation. Compound (**49**, m.p. 174°) was prepared
from (**25**) and crystallisation of the (-)-ω-camphanates of (**49**) gave the pure
camphanate [m.p. 200°, $[\alpha]_D$ -4°(CHCl$_3$)] of the chiral diol [**50**, m.p. 155°, $[\alpha]_D$ -14°
(CHCl$_3$)]. Similarily the racemic derivative (**54**, m.p. 176°), prepared from (**42**) was
resolved to give the chiral diol [**55**, m.p. 167°, $[\alpha]_D$ -10°(CHCl$_3$)].

   It was also found that benzylation of the dibutylstannylene derivative of the
tetra-O-benzyl ether (**47**) gave predominantly (*ca.* 85%) the penta-O-benzyl ether
(**48**) and this was separated from the regioisomer by chromatography. The other two
analogues (**50** and **55**) of (**47**) behaved similarily in the tin-mediated benzylation.
Thus from the chiral derivatives (**47, 50** and **55**) we were able to prepare the chiral
derivatives (**48, 52, 53, 21,10** and **57**) suitable for the synthesis of D-*myo*-inositol 4-,
1,3,4-tris-, 1,4-bis-, 1,3,4,5-tetrakis-, 1,4,5-tris- and 1,5-bis-phosphates respectively.
The latter phosphate is of interest as it has been shown (*29*) that the 1,5-bis-
phosphate of cyclohexane 1,3,5-triol mobilises calcium ions as efficiently as
D-*myo*-inositol 1,4,5-trisphosphate in *Neurospora crassa*. The 2,3,5,6-tetra-O-
-benzyl-D-*myo*-inositol (**53**) is also available by crystallisation of the bis-(+)-
-ω-camphanate of the racemic tetra-O-benzyl ether and likewise the enantiomer of
(**53**), 1,2,4,5-tetra-O-benzyl-D-*myo*-inositol, can be prepared by crystallisation of the
bis-(-)-ω-camphanate. The latter tetra-O-benzyl ether is a suitable intermediate for
the synthesis of D-*myo*-inositol 3,6-bisphosphate which is a recently (*30*) described
intermediate in inositol phosphate metabolism. The enantiomers of compounds (**48**,

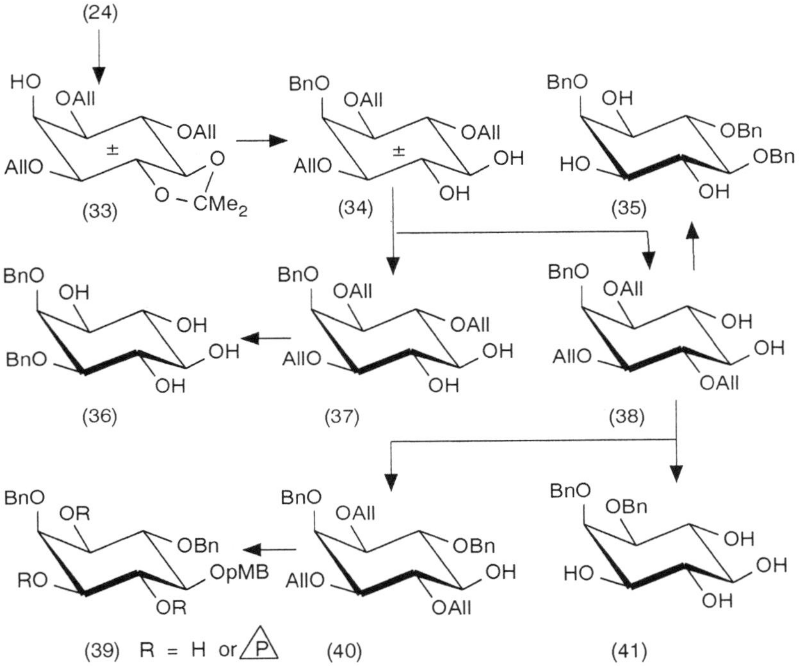
(24)
HO
OAll
OAll
±
AllO
O
O – CMe2
(33)
BnO
OAll
OAll
±
AllO
OH
OH
(34)
BnO
OH
OBn
HO
OBn
OH
(35)
BnO
OH
OH
BnO
OH
OH
(36)
BnO
OAll
OAll
AllO
OH
OH
(37)
BnO
OAll
OH
AllO
OH
OAll
(38)
BnO
OR
OBn
RO
OpMB
OR
(39) R = H or P
BnO
OAll
OBn
AllO
OH
OAll
(40)
BnO
OBn
OH
HO
OH
OH
(41)

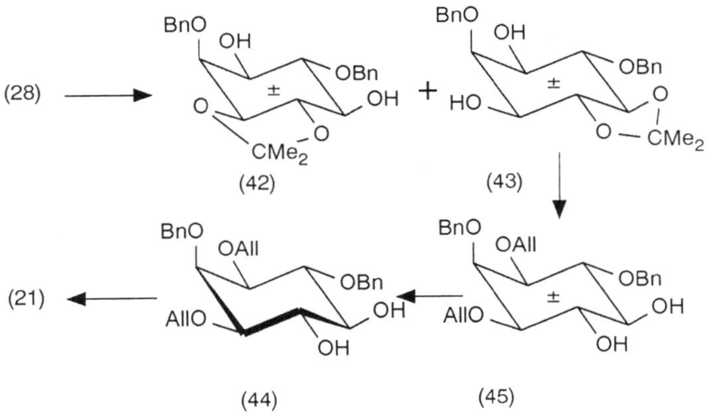
(28)
BnO
OH
OBn
±
O
OH
O
CMe2
(42)
+
BnO
OH
OBn
±
HO
O
O — CMe2
(43)
(21)
BnO
OAll
OBn
AllO
OH
OH
(44)
BnO
OAll
OBn
±
AllO
OH
OH
(45)

52, 53, 21, 10 and 57) were also available from the enantiomers of compounds (47, 50 and 55) prepared using (+)-ω-camphanic acid for the resolution.

Benzylation of the symmetrical tetra-O-allyl ether (27) and subsequent deallylation using Pd/C gave the highly crystalline 2,5-di-O-benzyl-*myo*-inositol (29, m.p. 272°). In contrast to the preparation of compound (28) the yields in this reaction were consistently high because the product crystallised from the reaction mixture. Because of the low solubility of (29) in organic solvents we considered that it might not be a suitable intermediate for phosphorylation studies and therefore prepared from it the isopropylidene derivative (31, m.p. 186°). Both the tetrol (29) and the isopropylidene derivative (31) were subsequently phosphorylated successfully (see below). *p*-Methoxybenzylation of the tetra-O-allyl ether (27) and subsequent deallylation gave 2,5-di-O-*p*-methoxybenzyl-*myo*-inositol (30, m.p.230°) which also gave a mono-isopropylidene derivative (32, m.p. 143°). This was benzylated and deacetonated to give the diol (58, m.p. 158°). Crystallisation of the bis-(-)-ω-camphanates of (58) gave the pure camphanate of the chiral diol (59) which gave predominantly the benzyl ether (61) on tin-mediated benzylation. Removal of the *p*-methoxybenzyl group from (61) gave 1,3,6-tri-O-benzyl-D-*myo*--inositol (60) suitable as an intermediate for the synthesis of D-*myo*-inositol 2,4,5-trisphosphate which has found use in many experiments (*27,31*).

Intermediates for the Synthesis of Phosphatidylinositol Phosphates.

The chiral intermediates described above are also suitable for the synthesis of all of the known phosphatidylinositol phosphates. Saponification of compound (15) and subsequent allylation, deacetonation, benzylation and deallylation will give (62) which is an intermediate for the synthesis of phosphatidylinositol 3-phosphate. This recently discovered lipid and the related 3,4-bis- and 3,4,5-trisphosphates are of interest since it appears that their metabolism may be regulated by growth factor receptor and oncogene-encoded protein tyrosine kinases (*32-37*).

Phosphorylation of compound (50) will give (63) and removal of *p*-methoxybenzyl group will give the intermediate (64) suitable for the synthesis of phosphatidylinositol 3,4-bisphosphate. Similarily compound (51) will give (65) on phosphorylation and after removal of the *p*-methoxybenzyl group the intermediate (66) is suitable for the synthesis of phosphatidylinositol 4-phosphate. Saponification of the camphanate (13) gives (67) which can be converted into (68). Phosphorylation of (68) and removal of the *p*-methoxybenzyl group will give (69) suitable for the synthesis of phosphatidylinositol 4,5-bisphosphate.

Compound (70), which is the enantiomer of (55), can be prepared from the racemate (54) by preferential crystallisation of the bis-(+)-ω-camphanate. Tin-mediated allylation of (70) will give (71) and this after benzylation and subsequent deallylation will give (72) a suitable intermediate for the synthesis of phosphatidylinositol 3,5-bisphosphate. We have shown that partial de-*p*-methoxybenzylation of (19) and (55) gives the triol (22) in reasonable yield and it is readily separable from other products of the reaction. We have converted both the triols (18 and 22) into the trisphosphate (73) and this is being used as an intermediate for the synthesis of phosphatidylinositol 3,4,5-trisphosphate a compound which may also be involved in the modulation of F-actin polymerisation (*38*).

Intermediates for the Synthesis of the 'Lipid Anchor' of Cell Surface Glycoproteins and the Phosphatidylinositol Mannosides of *Mycobacterium tuberculosis*.

We have recently described (*6*) in detail the synthesis of the chiral intermediate (74) and its derivatives required for these syntheses. One of the intermediates used for

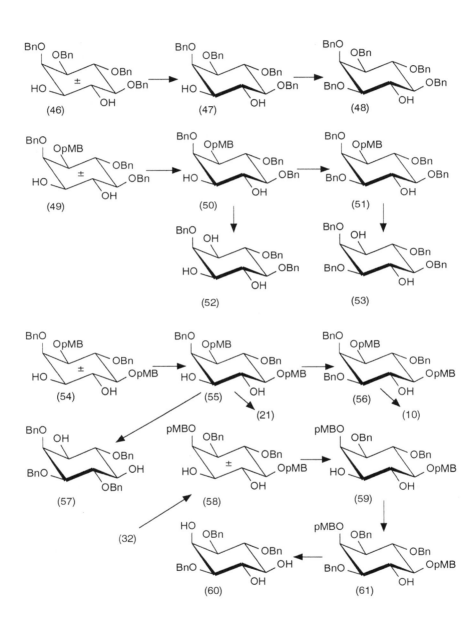

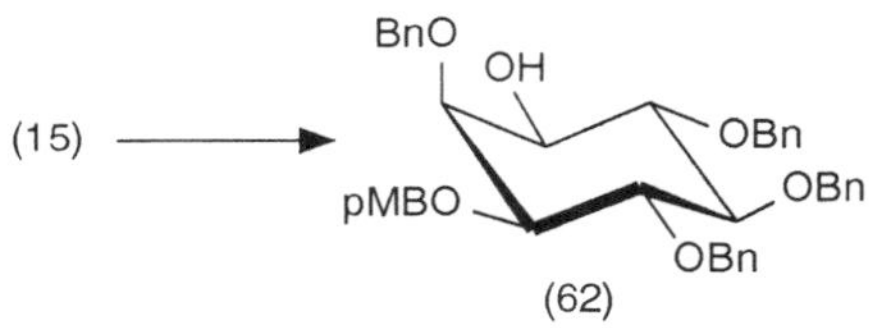

(15) ⟶

BnO OH
OBn
pMBO OBn
OBn
(62)

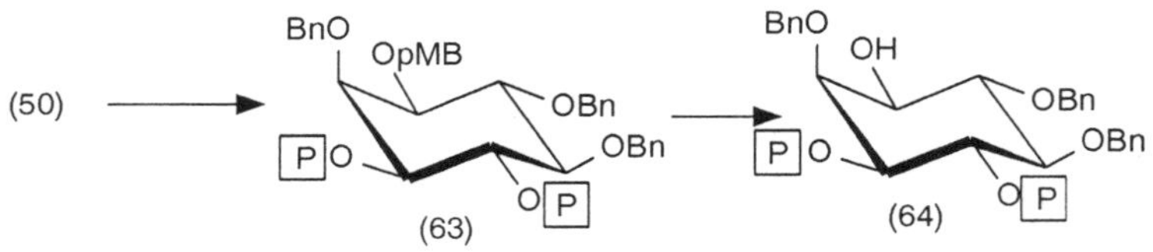

(50) ⟶

BnO OpMB
OBn
[P]O OBn
O[P]
(63)

⟶

BnO OH
OBn
[P]O OBn
O[P]
(64)

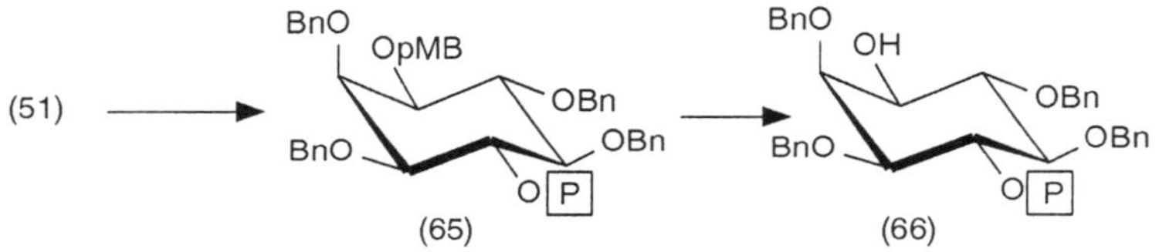

(51) ⟶

BnO OpMB
OBn
BnO OBn
O[P]
(65)

⟶

BnO OH
OBn
BnO OBn
O[P]
(66)

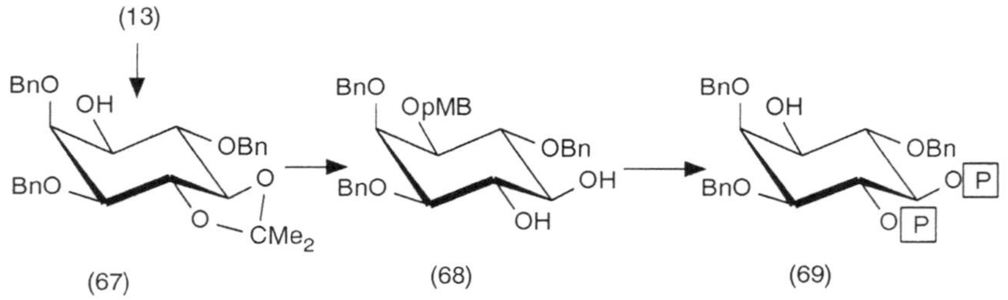

(13)
↓

BnO OH
OBn
BnO O
O—CMe₂
(67)

⟶

BnO OpMB
OBn
BnO OH
OH
(68)

⟶

BnO OH
OBn
BnO O[P]
O[P]
(69)

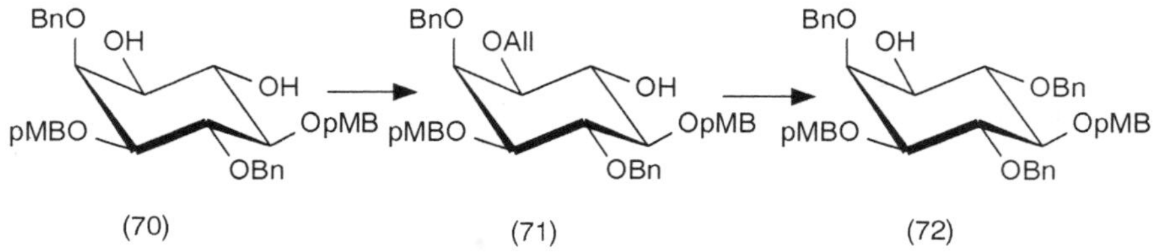

BnO OH
OH
pMBO OpMB
OBn
(70)

⟶

BnO OAll
OH
pMBO OpMB
OBn
(71)

⟶

BnO OH
OBn
pMBO OpMB
OBn
(72)

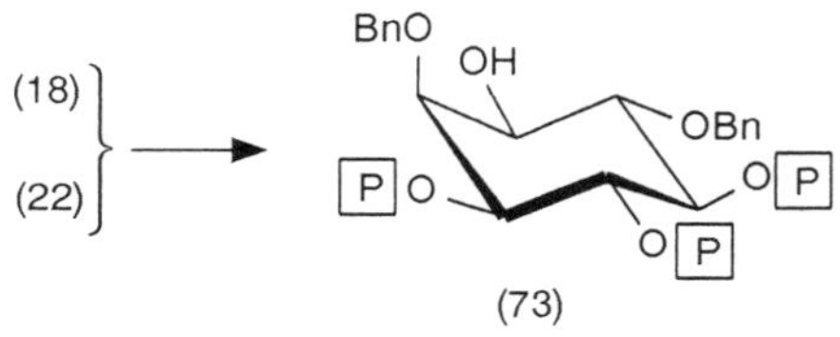

(18)
(22) } ⟶

BnO OH
OBn
[P]O O[P]
O[P]
(73)

the preparation of (74) was the racemate (75) (6) and this on tin-mediated benzylation gave (76) which was readily resolved by crystallisation of the ω-camphanates. The chiral derivative [79, m.p. 85°, $[\alpha]_D$ -26°(CHCl$_3$), obtained by saponification of the pure (-)-ω-camphanate] was converted by a series of standard reactions into the bis-crotyl ether (78) which on decrotylation gave (77) a suitable intermediate for the synthesis of D-myo-inositol 1,4-bisphosphate. De-p-methoxy-benzylation of (77) will give 2,3,5-tri-O-benzyl-D-myo-inositol (80) a suitable intermediate for the synthesis of D-myo-inositol 1,4,6-trisphosphate. The enantiomer of (79) [obtained by saponification of the pure (+)-ω-camphanate of (76)] can be similarly converted to the enantiomers of compounds (77) and (80) which are suitable intermediates for the synthesis of D-myo-inositol 3,6-bis- and 3,4,6-trisphosphates respectively. The 1,4,6- and 3,4,6-trisphosphates are recently described intermediates in the metabolism of myo-inositol 1,3,4,6-tetrakisphosphate in avian erythrocytes (20) and in *Dictyostelium* (30).

Phosphorylation Studies.

Most of the phosphorylations which we have carried out in Mill Hill have been with the phosphoramidite reagents (81 and 82), prepared as described previously (39) from dichloro-(N,N-diisopropylamino)phosphine (40,41) and some examples are given here.

The diol (83) [prepared from compound (11)] was converted into the bisphosphate [84, m.p. 86°, $[\alpha]_D$ +4°(CHCl$_3$)] and the p-methoxybenzyl group was removed to give [85, syrup, $[\alpha]_D$ +5°(CHCl$_3$)]. Further phosphorylation of (85) gave the protected trisphosphate [86, m.p.105°, $[\alpha]_D$ +5°(CHCl$_3$)] which was also obtained by direct phosphorylation of (10) with the phosphoramidite (82). Deprotection of (86) gave D-myo-inositol 1,4,5-trisphosphate. The enantiomer of (86) was also prepared *via* the camphanate (4). The L-myo-inositol 1,4,5-tris-phosphate derived from this does not mobilise calcium ions (42).

Phosphorylation of the isopropylidene derivative (31) and deacetonation of the product (87) gave the diol (88) which was phosphorylated to give the crystalline protected tetrakisphosphate [89, m.p. 109°]. This was also obtained by direct phosphorylation of 2,5-di-O-benzyl-myo-inositol (29). Deprotection of (89) gave myo-inositol 1,3,4,6-tetrakisphosphate and this synthetic sample has recently been shown to mobilise calcium ions in *Xenopus* oocytes (Ivorra,I.; Gigg,R.; Irvine,R.F. and Parker,I., Biochem. J.,(in press). The 3,4,5-tri-O-benzyl-1,2-O-isopropylidene--D-myo-inositol (74) was hydrolysed to the triol (90) and this on phosphorylation gave the protected trisphosphate (91) which on deprotection gave D-myo-inositol 1,2,6-trisphosphate (93) identical with the material prepared by phytase hydrolysis of inositol hexakisphosphate ('phytic acid'). The enantiomer of (93) was prepared similarily from the enantiomer of (74). Membrane permeable derivatives of the trisphosphate (93) which can be hydrolysed by enzymes are being prepared *via* (92).

Racemic 1,2,3,6-tetra-O-benzyl-myo-inositol (94) gave the protected bis--phosphate [95, m.p. 105°, cf also (43)] and phosphorylation of the racemate (3) and subsequent hydrolysis of the isopropylidene group gave the racemic phosphate (96, m.p. 124°). Phosphorylation of (96) with the phophoramidite (82) gave the racemate of the protected trisphosphate (86) as a syrup [cf also (44)]. Racemic 1,2,4,5-tetra--O-benzyl-myo-inositol gave the bis-phosphate (98, m.p. 98°). The racemate of (17, R = P̄ , m.p.98°) has been synthesised and converted into the racemate of (20, m.p.103°) as model compounds for the use of (20) in the synthesis of the 3-phosphorothioate analogue of D-myo-inositol 1,3,4,5-tetrakisphosphate and possibly for the synthesis of ^{3}H-labelled D-myo-inositol 1,4,5-trisphosphate *via* the ketone derived from (20).

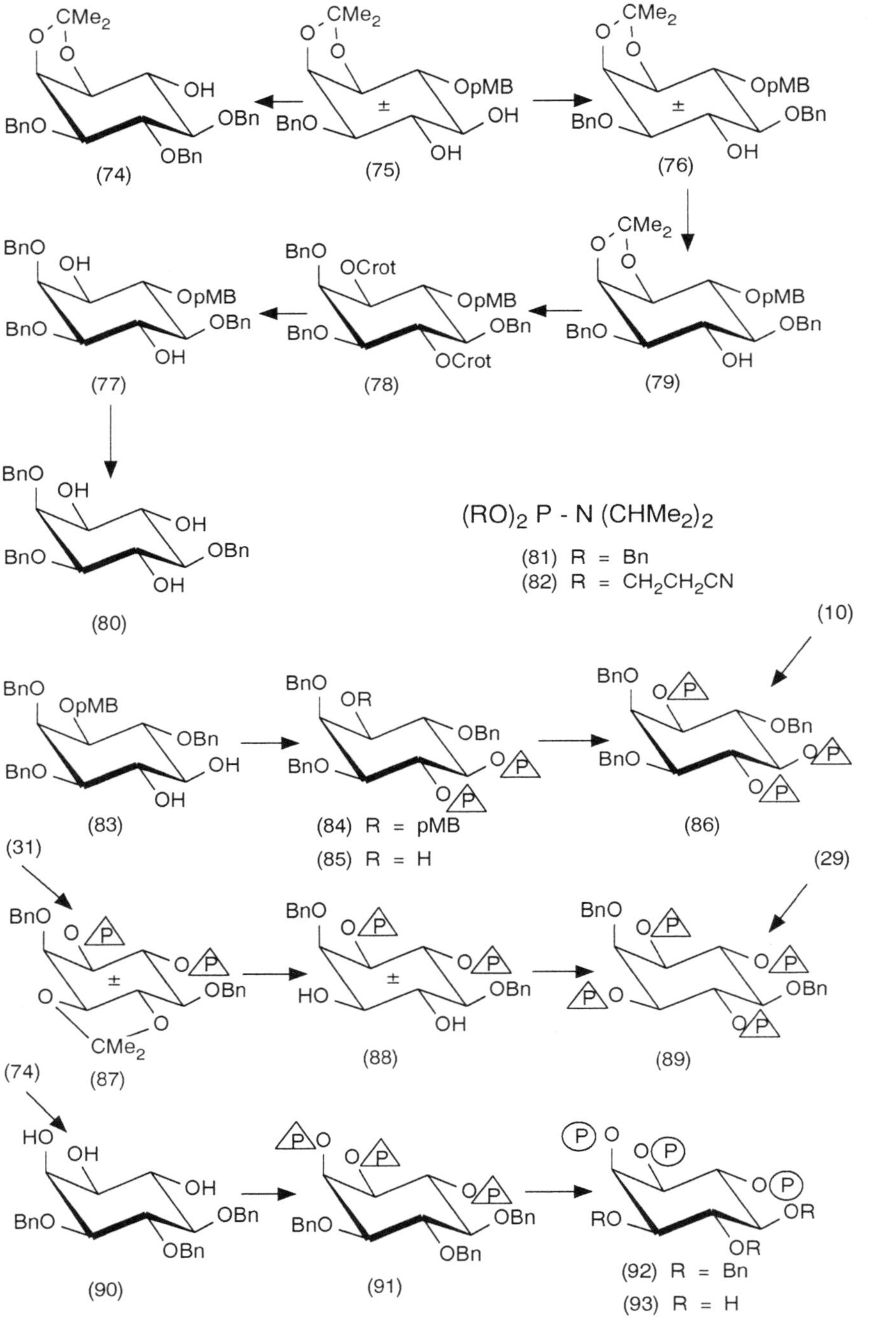
(74)
(75)
(76)
(77)
(78)
(79)
(80)
(RO)2 P - N (CHMe2)2
(81) R = Bn
(82) R = CH2CH2CN
(10)
(83)
(84) R = pMB
(85) R = H
(86)
(31)
(29)
(87)
(88)
(89)
(74)
(90)
(91)
(92) R = Bn
(93) R = H

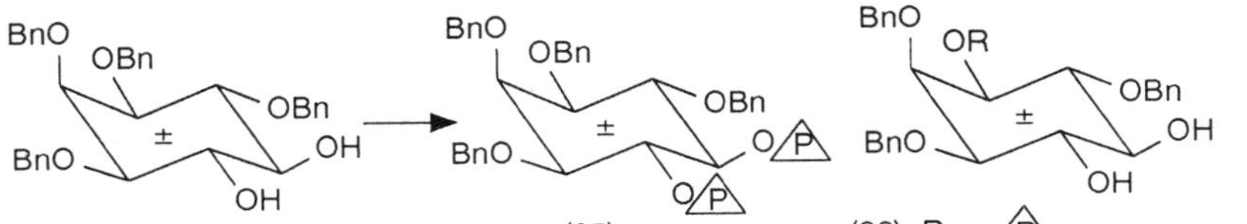

(94)

(95)

(96) R = △P
(97) R = PO(OCH₂CCl₃)₂

(98)

(99) R = PO(OCH₂CCl₃)₂

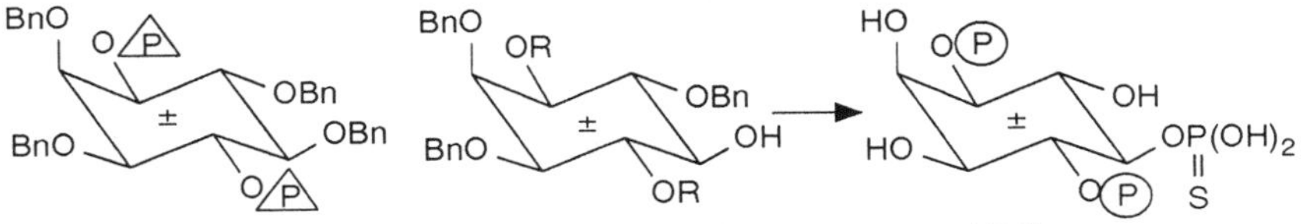

(100)

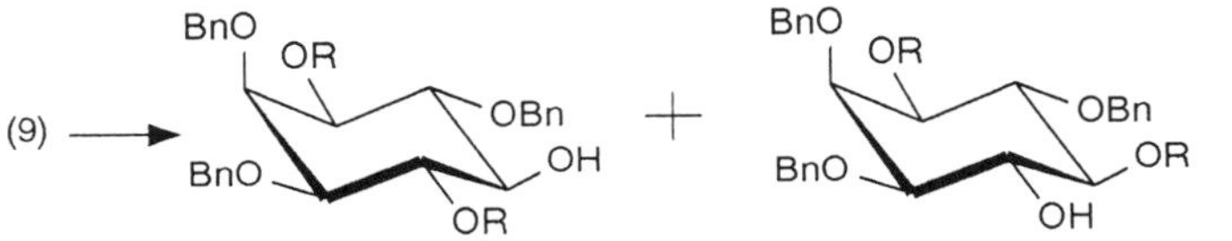

(9) ⟶

(101) R = All
(102) R = Prop

(103) R = All
(104) R = Prop

(105) R = All
(106) R = Prop

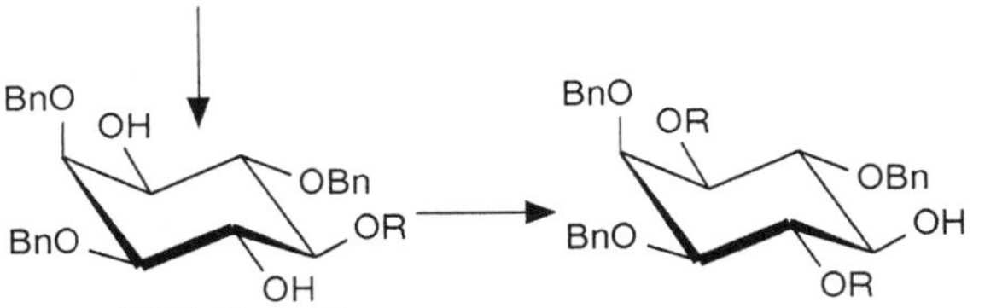

(107) R = PO(OCH₂CCl₃)₂

(108) R¹ = R² = H
(109) R¹ = All ; R² = H

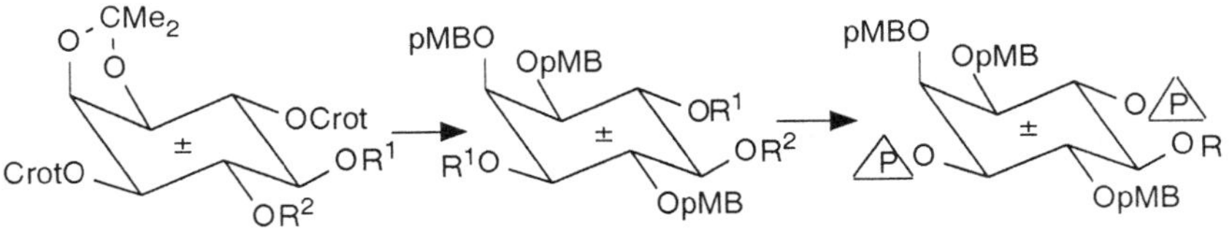

(110) R¹ = Crot ; R² = All
(111) R¹ = H ; R² = Prop

(112) R = Prop
(113) R = P(OCH₂CH₂CN)(OCHPh(oNO₂))=O with Me

Some of our early phosphorylation studies were carried out with bis-(trichloroethyl)-phosphorochloridate and the racemic phosphate (97) was prepared (45) from (3). Only one of the hydroxyl groups of (97) could be further phosphorylated with the phosphorochloridate but the product (99) was further phosphorylated (in collaboration with Prof. B.V.L. Potter) using phosphoramidite chemistry to give, after deprotection the 5-phosphorothioate analogue (100) of racemic myo-inositol 1,4,5-trisphosphate (45). We were able to phosphorylate (99) with phosphorus oxychloride and subsequent treatment with methanol gave a crystalline dimethylphosphate ester of (99). The tris-phosphorothioate analogue of racemic myo-inositol 1,4,5-trisphosphate was prepared from the racemate of (10) (46,47). These phosphorothioate analogues mobilise calcium ions like the phosphates but are stable to the phosphatases which metabolise D-myo-inositol 1,4,5-trisphosphate (48).

In order to make the chiral analogue of the phosphorothioate (100) we shall carry out a series of reactions on the chiral allyl ether (9) which we have previously studied with the racemate (7). Tin-mediated allylation of (9) gives an inseparable mixture of the bis-allyl ethers (101 and 103) but after isomerisation of the allyl groups to prop-1-enyl groups [using potassium t-butoxide in dimethyl sulphoxide (11)] the bis-prop-1-enyl ethers (102 and 104) are readily separated by chromatography (7). Allylation of (102) and subsequent acidic hydrolysis of the prop-1-enyl groups gives the allyl ether (105) which is isomerised to the prop-1-enyl ether (106). Phosphorylation of (106) with bis-(trichloroethyl)-phosphorochloridate and subsequent removal of the prop-1-enyl group with dilute acid gives (107) which is the chiral analogue of (99).

Related intermediates are being used to prepare a 'caged' derivative of myo--inositol 1,4,5-trisphosphate. Racemic 1,4-di-O-(but-2-enyl)-2,3-O-isopropylidene--myo-inositol (108, m.p. 105°) was converted by tin-mediated allylation into the 5-O-allyl ether (109) which on deacetonation and subsequent p-methoxybenzylation gave (110, m.p. 92°). Treatment of the latter with potassium t-butoxide in dimethyl sulphoxide (11) gave 1,2,4-tri-O-p-methoxybenzyl-5-O-(prop-1-enyl)-myo-inositol (111,m.p. 123°). This will be converted into the bisphosphate (112) and, after removal of the prop-1-enyl group, further phosphorylation using 2-cyanoethyl N,N-diisopropylchlorophosphoramidite will give the protected trisphosphate (113). Basic hydrolysis of the 2-cyanoethyl groups and subsequent oxidative removal of the p-methoxybenzyl groups should give myo-inositol 1,4,5-trisphosphate with a 'cage' on the 5-phosphate.

Conclusion.

We have described several derivatives of myo-inositol which can be resolved readily and in quantity by crystallisation of the (+)- and/or (-)-ω-camphanates. These chiral derivatives are suitable for the synthesis of most of the inositol phosphates of the 'phosphatidylinositol cycle' and all of the known phosphatidylinositol phosphates as well as the basic lipid structure of the 'lipid anchor' of cell membrane glycoproteins, the phosphatidylinositol mannosides of Mycobacterium tuberculosis and of the plant lipid 'phytoglycolipid'.

Literature Cited.

1.   Gigg, J.; Gigg, R.; Payne, S. and Conant, R. Carbohydr. Res. 1985, 140, C1-C3.
2.   Gigg, J.; Gigg, R.; Payne, S. and Conant, R. Carbohydr. Res. 1985, 142, 132-134.

3. Gigg, J.; Gigg, R.; Payne, S. and Conant, R. *J. Chem. Soc., Perkin Trans. 1* **1987**, 423-429.
4. Gigg, J.; Gigg, R.; Payne, S. and Conant, R. *J. Chem. Soc., Perkin Trans. 1* **1987**, 1757-1762.
5. Gigg, J.; Gigg, R.; Payne, S. and Conant, R. *J. Chem. Soc., Perkin Trans. 1* **1987**, 2411-2414.
6. Desai, T.; Fernandez-Mayoralas, A.; Gigg, J.; Gigg, R. and Payne, S. *Carbohydr. Res.* **1990**, *205*, 105-123.
7. Gigg, R.H., International Patent Application No. WO 89/00156 (12 Jan. 1989 to 3i Research Exploitation Ltd. formerly Research Corporation Ltd.) *Chem. Abstr.* **1989**, *111*, 174590t.
8. Desai, T. *The Chemistry of Inositol Compounds of Biological Interest*; M. Phil. Thesis, C.N.A.A. **1989**.
9. Potter, B.V.L. *Nat. Prod. Reports* **1990**, 1-24.
10. Billington, D.C. *Chem. Soc. Rev.* **1989**, *18*, 83-122.
11. Gigg, R. *ACS Symp. Ser.* **1977**, *39*, 253-278.
12. Jacques, J.; Collet, A. and Wilen, S.H. *Enantiomers, Racemates and Resolutions*; J. Wiley: New York, **1981**.
13. Wilen, S.H.; Collet, A. and Jacques, J. *Tetrahedron* **1977**, *33*, 2725-2736.
14. Vieira, E. and Vogel, P. *Helv. Chim. Acta* **1983**, *66*, 1865-1871.
15. Wagner, J.; Vieira, E. and Vogel, P. *Helv. Chim. Acta* **1988**, *71*, 624-630.
16. Angyal, S.J.; Randall, M.H. and Tate, M.E. *J. Chem. Soc. C* **1967**, 919-922.
17. Shears, S.B. *Biochem. J.* **1989**, *260*, 313-324.
18. Balla, T.; Hunyady, L.; Baukal, A.J. and Catt, K.J. *J. Biol. Chem.* **1989**, *264*, 9386-9390.
19. Menniti, F.S.; Oliver, K.G.; Nogimori, K.; Obie, J.F.; Shears, S.B. and Putney, J.W. *J. Biol. Chem.* **1990**, *265*, 11167-11176.
20. Stephens, L.R. and Downes, C.P. *Biochem. J.* **1990**, *265*, 435-452.
21. Ozaki, S.; Kondo. Y.; Nakahira, H.; Yamaoko, S. and Watanabe, Y. *Tetrahedron Lett.* **1987**, *28*, 4691-4694.
22. Baudin, G.; Glänzer, B.I.; Swaminathan, K.S. and Vasella, A. *Helv. Chim. Acta* **1988**, *71*, 1367-1378.
23. Dreef, C.E.; Tuinman, R.J.; Elie, C.J.J.; van der Marel, G.A. and van Boom, J.H. *Recl. Trav. Chim. Pays-Bas* **1988**, *107*, 395-397.
24. Watanabe. Y.; Oka, A.; Shimizu, Y. and Ozaki, S. *Tetrahedron Lett.* **1990**, *31*, 2613-2616.
25. Cullen, P.J.; Irvine, R.F.; Drobak, B.K. and Dawson, A.P. *Biochem. J.* **1989**, *259*, 931-933.
26. Berridge, M.J. and Irvine, R.F. *Nature (London)* **1989**, *341*, 197-205.
27. Boynton, A.L.; Dean, N.M. and Hill, T.D. *Biochem. Pharmacol.* **1990**, *40*, 1933-1939.
28. Boss, R. and Scheffold, R. *Angew. Chem., Int. Ed. Engl.* **1976**, *15*, 558-559.
29. Schultz, C.; Gebauer, G.; Metschies, T.; Rensing, L. and Jastorff, B. *Biochem. Biophys. Res. Commun.* **1990**, *166*, 1319-1327.
30. Stephens, L.R. and Irvine, R.F. *Nature (London)* **1990**, *346*, 580-583.
31. Cullen, P.J.; Irvine, R.F. and Dawson, A.P. *Biochem. J.* **1990**, *271*, 549-553.
32. Serunian, L.A.; Auger, K.R.; Roberts, T.M. and Cantley, L.C. *J. Virol.* **1990**, *64*, 4718-4725.
33. Pignataro, O.P. and Ascoli, M. *J. Biol. Chem.* **1990**, *265*, 1718-1723.
34. Sultan, C.; Breton, M.; Mauco, G.; Grondin, P.; Plantavid, M. and Chap, H. *Biochem. J.* **1990**, *269*, 831-834.
35. Kucera, G.L. and Rittenhouse, S.E. *J. Biol. Chem.* **1990**, *265*, 5345-5348.
36. Nolan, R.D. and Lapetina, E.G. *J. Biol. Chem.* **1990**, *265*, 2441-2445.
37. Downes, C.P. and Macphee, C.H. *Eur. J. Biochem.* **1990**, *193*, 1-18.

38.   Eberle, M.; Traynor-Kaplan, A.E.; Sklar, L.A. and Norgauer, J. *J. Biol. Chem.* **1990**, *265*, 16725-16728.
39.   Bannwarth, W. and Trzeciak, A. *Helv. Chim. Acta* **1987**, *70*, 175-186.
40.   Tanaka, T.; Tamatsukuri, S. and Ikehara, M. *Tetrahedron Lett.* **1986**, *27*, 199-202.
41.   Damha, M.J. and Ogilvie, K.K. *J. Org. Chem.* **1986**, *51*, 3559-3560.
42.   Strupish, J.; Cooke, A.M.; Potter, B.V.L.; Gigg, R. and Nahorski, S.R. *Biochem. J.* **1988**, *253*, 901-905.
43.   Hamblin, M.R.; Potter, B.V.L. and Gigg, R. *J. Chem. Soc., Chem. Commun.* **1987**, 626-627.
44.   Cooke, A.M.; Potter, B.V.L. and Gigg, R. *Tetrahedron Lett.* **1987**, *28*, 2305-2308.
45.   Cooke, A.M.; Noble, N.J.; Payne, S.; Gigg, R. and Potter, B.V.L. *J. Chem. Soc., Chem. Commun.* **1989**, 269-271.
46.   Cooke, A.M.; Gigg, R. and Potter, B.V.L. *J. Chem. Soc., Chem. Commun.* **1987**, 1525-1526.
47.   Potter, B.V.L., International Patent Application No. WO 88/07047 (22 Sept. 1988 to 3i Research Exploitation Ltd. formerly Research Corporation Ltd.); *Chem. Abstr.* **1989**, *111*, 39827k.
48.   Wojcikiewicz, R.J.H.; Cooke, A.M.; Potter, B.V.L. and Nahorski, S.R. *Eur. J. Biochem.* **1990**, *192*, 459-467.

RECEIVED February 11, 1991

Chapter 7

Pharmacological Effects of D-*myo*-Inositol-1,2,6-Trisphosphate

M. Sirén[1], L. Linné[2], and L. Persson[2,3]

[1]Department of Internal Medicine, Academic Hospital, Uppsala, Sweden
[2]Perstorp Pharma, S-284 80 Perstorp, Sweden

PP56, D-myo-inositol-1,2,6-trisphosphate (1,2,6-IP$_3$) is a new chemical entity for therapeutic use developed by Perstorp Pharma, Sweden, in collaboration with the originator Ass. Prof. M.J. Sirén, Switzerland (M.J. Sirén, US Patent No. 4735 936, EPO Patent No. 179 439, 1984).

The physiological role of different inositol phosphates is presently undergoing extensive research. The effects observed with different analogues seem to be very specific. The most studied analogue is the intracellular second messenger 1,4,5-IP$_3$, which stimulates intracellular release of calcium affecting cellular metabolism and secretion. In contrast, PP56 does not induce calcium release (1).

PP56 is manufactured via enzymatic degradation of phytic acid and purified by a chromatographic process.

Several pharmacological effects of PP56 have been documented in animal models representing acute as well as chronic diseases. In acute inflammatory models PP56 has been shown to counteract physical, immunological as well as agonist-induced inflammation. For example in a thermal skin injury model, intravenous administration of PP56 causes a significant decrease in protein extravasation. Furthermore, the compound counteracts the inflammation and oedema induced by carrageenan in a foot pad model. In chronic inflammatory models, such as adjuvant arthritis, the administration of PP56 significantly reduces the inflammatory response.
    PP56 has also been studied in connection with prevention of secondary diabetic complications in diabetic rats. The administration of the compound results in significant improvements in nerve conduction velocity and a reduction in the formation of ketone bodies, reduction of lens cataracts and also a total reduction of mortality.

[3]Corresponding author

0097–6156/91/0463–0103$06.00/0

A primary pharmacological characteristic of the substance is to inhibit vascular leakage and oedema formation. One proposed mechanism of action of the compound is to protect the endothelial layer against degradation via different mediators and cell types. Preliminary observations have shown that PP56 counteracts some receptor mediated interactions which are involved in these conditions. Investigation of the interaction of PP56 and different vasoactive agents shows that the compound in low concentrations selectively inhibits the effects of neuropeptide Y, a potent vasoconstrictor. Furthermore, it has been shown that platelet aggregability and thrombus formation are inhibited. The compound also suppresses oxidative processes such as lipid peroxidation.

Purification and Structure

Phytic acid is the starting material in the manufacturing process of PP56. Degradation of phytic acid can be performed chemically or enzymatically resulting in the formation of mixtures of isomers of inositol phosphates. For the preparation of some specific isomers enzymatic degradation has been utilized before. However, for the preparation of pure D-myo-inositol-1,2,6-trisphosphate a specific phytase from yeast has to be used. The enzymatic source is suspended in water and the pH, temperature and hydrolysis time is chosen to produce maximum amounts of PP56. The purification of PP56 from minor amounts of other inositol phosphates, proteins, buffer etc is performed by the utilization of ion exchange chromatography with gradient elution. For intravenous applications a pentasodium hydrogen salt of PP56 is prepared after neutralization and precipitation.

The chemical structure of PP56 is shown in Figure 1 (2). Two optical isomers of myo-inositol-1,2,6-trisphosphate exist. The pure L- and D-forms were prepared via a synthetic route which utilized a C-3 - C-5 protected myo-inositol derivative. This racemic mixture was converted to diastereomeric derivatives which were separated by column chromatography. The obtained enantiomers were phosphorylated and deprotected resulting in pure enantiomeric forms of inositol trisphosphate (3).

A comparison of the optical rotation of these enantiomers with PP56 produced enzymatically shows the latter to be D-myo-inositol-1,2,6-trisphosphate.

Anti-inflammatory Pharmacological Effects

In acute inflammatory models PP56 has been shown to counteract different types of inflammations. Common to all these models is that they cause vascular leakage of proteins leading to the formation of oedema.

In a thermal skin injury model, intravenous administration of PP56 causes a dramatic decrease in protein extravasation. Burn injury was induced in anesthetized rats by exposing the abdominal skin to a temperature of 55°C by means of a hot plate. A continuous registration of the temperature was performed and the skin exposure was interrupted when the temperature had decreased to 45°C. By this procedure a full-thickness burn trauma of the skin was induced as judged from histological sections. Fifteen minutes after the burn injury, PP56 was given as a bolus dose followed by a continuous intravenous infusion during two hours. Evans blue (EB) was injected i.v. 90 minutes after the start of the infusion. Thirty minutes after the administration of EB, the burned skin area was dissected and the extravasation of EB-bound albumin was quantified by a spectrophotometric technique. The experimental result shows that PP56 in a dose-dependant way significantly reduces the vascular leakage without any side effects. Earlier experience with this model indicates that a similar magnitude of response cannot be achieved by non steroidal anti-inflammatory drugs (NSAID) or corticosteroids.

The efficacy of PP56 was also assessed in a rat model of acute inflammation, i.e. foot pad oedema . Rats were given either an i.v. bolus followed by i.v. infusion of PP56 or saline vehicle as control. Acute inflammation was initiated in one hind-paw using a single subplantar injection of 0.1 ml 1 % w/v carrageenan in sterile normal saline. The inflammation was assessed hourly for 5 hours and finally at 24 hours by measurement of paw circumference immediately proximal to the two hindmost spurs of the foot . In figure 2 it is shown that a bolus of PP56 followed by infusion significantly reduces the oedema by approximately 50 % 2-4 hours after the injection of carrageenan.

In another experimental model adjuvant arthritis was induced in rats and the progress of the inflammation was compared when PP56 or saline were administered (4). Daily dosing of PP56 showed marked suppression in inflammation compared to control. Moreover radiographic analysis of the hind-limbs of the rats showed a reduction in bone erosion when PP56 was administered.

Secondary Complications of Diabetes

The effects of PP56 were investigated in studies in rat with streptozotocin-induced diabetes. In one series of experiments (5) PP56 was mixed in the diet. The effect of the compound was to completely normalize the platelet reactivity against ADP and thrombin. The treatment also reduced the amount of urinary ketone bodies. Furthermore, there was observed a reduced mortality rate in the animals treated with PP56.

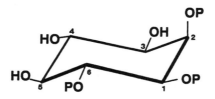

$$P = PO_3R_2, \text{ where } R = 5 \text{ Na, 1H}$$

Figure 1. Structure of PP56.

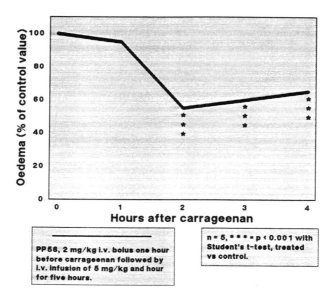

Figure 2. Footpad oedema in rats (intravenous infusion of PP56).

In another series of experiments the effect of PP56 on nerve conduction velocity in diabetic rats were investigated (6). Streptozotocin-induced diabetic rats were treated with PP56 (1 mg/h s.c; Alzet osmotic pump 2ML4) and were compared to one control group without diabetes and one diabetic group not receiving PP56. The treatment was maintained for three weeks. Motor nerve conduction velocity (MNCV) was then measured under halothane anaesthesia in the sciatic/tiblis system. MNCV in diabetic rats were reduced compared to controls. The treatment with PP56 significantly improved the nerve conduction velocity which can be seen in figure 3. Other experiments in this area indicate that the effects of PP56 are mediated via a different mode of action compared to aldose reductase inhibitors.

Neuropeptide Y Antagonistic Properties

In one set of experiments the objective was to investigate the possible interaction between PP56 and different vasodilators and vasoconstrictors such as acetylcholine, calcitonin-gene related peptide, substance P, endothelin, neuropeptide Y (NPY), histamine and bradykinin (7). The basilar artery from the brain of guinea pigs were dissected free and cut into cylindrical segments. The segments were mounted in a way which allowed the force of the vascular tension to be measured. NPY induced a strong concentration-dependent contraction of the artery. At concentrations of 10^{-8} M to 10^{-6} M PP56 significantly decreased the NPY-induced contraction, figure 4. These data show that PP56 is the first non-peptide which potently and selectively antagonizes NPY-induced contraction.

Levels of NPY are elevated in cardiovascular diseases and are not reduced by currently available antihypertensive agents, a factor which has been suggested to contribute to the relatively poor clinical outcome in treated hypertension despite adequate control of blood pressure.

In addition to its direct potent vasoconstricting effects NPY potentiates the vasoconstriction activity of noradrenaline and inhibits some vasodilators.

PP56 also inhibits NPY's potentiating effect on noradrenaline. The compound has also been shown to inhibit NPY-induced hypertension in animal studies.

On-going and Future Development

The present toxicological documentation demonstrates that PP56 can be given to humans with very good safety. The tolerability also after intravenous administration is very good.

MNCV (m/s)

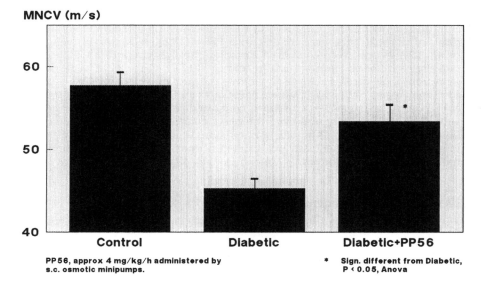

PP56, approx 4 mg/kg/h administered by   * Sign. different from Diabetic,
s.c. osmotic minipumps.   P < 0.05, Anova

Figure 3. Nerve conduction velocity in rats (parenteral PP56).

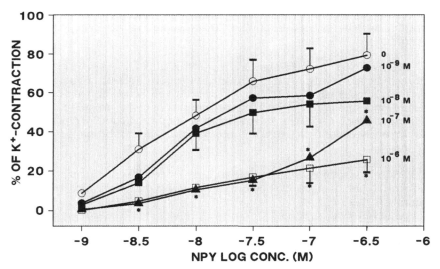

PP56 (10^{-8} - 10^{-6} M) caused progressive shift of NPY-induced contraction of
the guinea pig basilar artery.

Figure 4. Effect of PP56 on NPY-induced contraction.

Phase I clinical trials with PP56 are now underway and these have suggested that anti-inflammatory effects are also seen in humans. In an oedema model in human volunteers, i.v. administration of PP56 has shown a significant reduction in vascular permeability. Further clinical development will be directed towards indications such as burns but also other therapeutic areas where the component of vascular leakage is dominant. Clinical studies assessing the effect of the compound against NPY-related conditions are planned to commence in the near future.

PP56 has shown oral activity in animal studies. However, its bioavailability after oral administration in humans has to be increased. New formulations for the compound in order to improve this parameter are currently being investigated. These formulations include different prodrug approaches in order to make the compound more lipophilic and also galenic formulations of IP_3 per se in order to create an environment which facilitates the intestinal resorption.

Conclusions

The pharmacological profile of D-myo-inositol-1,2,6-trisphosphate (PP56) has been described. The compound is currently developed for therapeutic use by Perstorp Pharma, Sweden. The combination of a very clean safety profile and a new mode of action together with interesting pharmacological effects indicates that the substance may be applied in several pathological conditions.

The initial objective is to develop the compound for acute anti-inflammatory and anti-oedema indications in man. Areas such as burns, gastrointestinal leakage and cerebral oedema would be suitable for the use of PP56 as an intravenous infusion.

Furthermore, prophylactic or symptomatic treatment of chronic inflammations, cardiovascular diseases and secondary diabetic complications may be potential areas for the therapeutic use of PP56.

Literature Cited

1.   Authi, K.S.; Gustafsson, T.O.; Crawford, N. *Trombosis Haemostasis* **1989**, *62* ,250

2.   Johansson, C.; Kördel, J.; Drakenberg, T. *Carbohydr. Res.* **1990**, *207*, 35-41

3.   Desai, T.; Fernandez-Mayoralas, A.; Gigg, J.; Gigg , R.; Payne, S. *Carbohydrate Research* **1990**, *205*, 105

4.   Claxson, A.; Morris, C.; Blake, D.; Sirén, M.; Halliwell, B.;
      Gustafsson, T.; Löfkvist, B.; Bergelin, I. *Agents and Actions* **1990**,
      29, 1/2, 68

5.   Ruf, J.; Ciavatti, M.; Gustafsson, T.; Renaud, S. *Diabetes* (in press)

6.   Carrington, A.; Ettlinger, C.; Calcutt, N.; Tomlinson, D.
      Diabetologia 33 (supplement) **1990** A92

7.   Edvinsson, L.; Adamsson, M.; Jansen I. *Neuropeptides* **1990**, *17*,
      99

RECEIVED March 11, 1991

Chapter 8

Synthesis of New *myo*-Inositol Derivatives Containing Phosphate, Sulfate, and Sulfonamido Groups

Peter Westerduin and Constant A. A. van Boeckel

Organon Scientific Development Group, Akzo Pharma, P.O. Box 20, 5340 BH Oss, Netherlands

In this paper, we wish to report the synthesis of two new 1,4,5-tris-substituted *myo*-inositols isosteric to IP$_3$, i.e. the less negatively charged 1,4,5-trissulphate **31** and the neutral 1,4,5-tris-sulphonamido derivative **32**. Furthermore, the synthesis of three new inositol phosphates, i.e. D-*myo*-inositol 1,5-bisphosphate **22**, D-*myo*-inositol 3,5-bisphosphate **23** and rac. *myo*-inositol 4,5-cyclic phosphate **25** will be described. Additionally, the preliminary biological results obtained with the new compounds will be presented, together with the results obtained with two highly sulphated synthetic heparin fragments (i.e. compounds **33** and **34**).

Phosphoinositides (PI's) have recently emerged as compounds of considerable interest on account of their relationship with many cellular processes (*1-15*). At the moment, it is generally established that receptor-mediated hydrolysis of the PI phosphatidyl inositol 4,5-bisphosphate generates two second messengers, D-*myo*-inositol 1,4,5-trisphosphate (IP$_3$) and diacylglycerol (DAG). The water soluble IP$_3$ may be released intracellularly to effect a Ca^{++} efflux from intracellular stores (*16,17*), while the hydrophobic DAG remains in the plasma membrane to activate protein kinase C (PKC) (*2,18*). These two branches of the bifurcating PI signalling system (IP$_3$/Ca^{++} and DAG/PKC) are intimately involved in the regulation of many different physiological responses. The biological significance of the PI signalling system makes it an attractive target for drug design (*13, 19*) and accounts for the synthetic effort in the field of inositol phospholipids and inositol(poly)phosphates (*20-24*).

Since IP$_3$ is still the only inositol phosphate for which a clear-cut role in intracellular Ca^{++} homeostasis has been demonstrated (*25*), there is a considerable interest in the synthesis of IP$_3$ agonists and antagonists. Hopefully such compounds may find application as new pharmacotherapeuticals. However, the development of specific drugs interfering within the IP$_3$ signalling system of a certain target tissue is difficult because of the ubiquitous nature of IP$_3$. Nevertheless, it has been demonstrated that heterogeneity in the signal transduction systems may occur in different tissues and cells. This is not only examplified by the multiple forms of G-proteins (*26-28*), the various isoenzymes of protein kinase C (*29*), phospholipase

0097–6156/91/0463–0111$06.00/0

C (*30*) and phosphodiesterase (*31*), but also by the IP_3 binding protein itself. The heterogeneous distribution of the latter receptor (*32*), the (putative) tissue specificity in IP_3 mediated Ca^{++} release (*9, 10*) and the differences in cellular distribution of drugs in general, suggest that new pharmaceuticals may be developed to interfere with the action of IP_3.

In order to design IP_3 derivatives acting as intracellular Ca^{++} agonist or antagonist, it is mandatory to elucidate the structural requirements for high affinity towards the IP_3 active site. In this respect, mapping of the substrate specificity of the IP_3 receptor proteins should be performed systematically with modified analogues of the natural ligand. Several studies towards stucture-activity relationships indicated the vicinal phosphates (*33*) on positions 4 and 5 (D-configuration) to be essential for Ca^{++}-release. The Ca^{++} mobilizing activity of inositol 1:2-cyclic,4,5-trisphosphate (*34*) was much lower than IP_3 itself, while inositol 1,4-bisphosphate was inactive. Furthermore, several groups reported recently the synthesis of IP_3 analogues modified at positions 2, 3 or 6 (see Figure 1). The only IP_3 analogues exhibiting

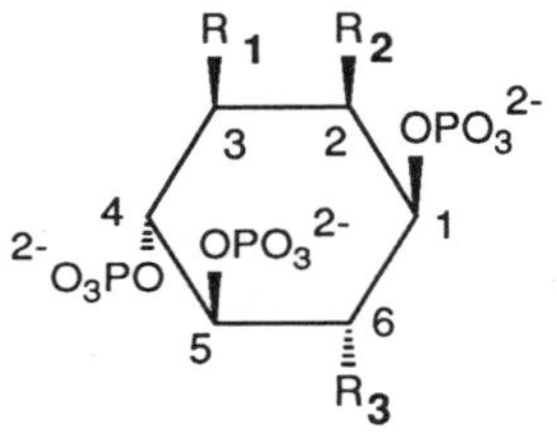

1	$R_1=R_3=OH$, $R_2=OAcyl$, $O(Alkyl)Aryl$ (*35*)
2	$R_1=F$, $R_2=R_3=OH$ (*36*)
3	$R_1=H$, $R_2=R_3=OH$ (*37*)
4	$R_1=R_2=R_3=H$ (*38*)
5	$R_1=R_2=OH$, $R_3=OCH_3$ (*39*), $R_3=H$, F, CH_3 (*40*)
6	$R_1=R_2=R_3=OButyryl$ (*41*)

Figure 1. IP_3 analogs at positions 2, 3, or 6.

considerable agonistic activity turned out to be the compounds modified at the 2-position (*35*), the 3-deoxy-3-fluoro analogue (*36*) and the 3-deoxy analogue (*37*) (**1, 2** and **3** resp., Figure 1).

However, not all combinations of possible inositol phosphates have been yet prepared. No biological data were available on inositol 1,5-bisphosphate and inositol derivatives containing a 4:5-cyclic phosphate, although it has been suggested that modification of the 4,5 locus would be profitable in order to design a molecule to antagonise Ca^{++} channel opening (*24*). On the other hand, the syntheses of two phosphorylated trihydroxycyclohexane derivatives, both interfering in the PI signalling system, were reported. Thus, the 3,5,6-trisdeoxy derivative of *myo*-inositol 1-monophosphate (**7**) was identified as a very potent inhibitor of

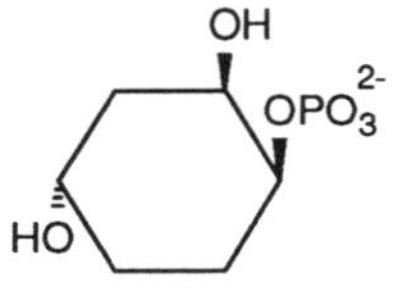

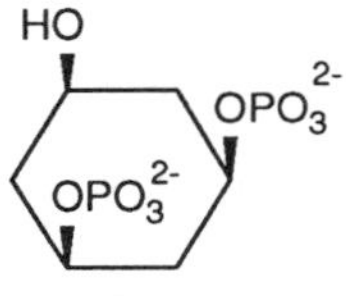

inositol monophosphatase (*42*), whereas cis,cis-cyclohexane 1,3,5-triol bisphosphate (**8**), a meso compound mimicking inositol 1,5-bisphosphate, turned out to be a full IP_3 agonist in releasing Ca^{++} from isolated vacuoles of Neurospora crassa (*43*). Since D-*myo*-inositol 1,5-bisphosphate, D-*myo*-inositol 3,5-bisphosphate and rac. *myo*-inositol 4:5-cyclic monophosphate, three isomers not (yet) detected in mammalian cells, have been uninvestigated thus far, synthesis and biological evaluation of these compounds in mammalian cell systems may be of interest. In this paper we decribe the synthesis of these new inositol phosphates [i.e. D-*myo*-inositol 1,5-bisphosphate (**22**) and 3,5-bisphosphate (**24**) (**44**) and rac. *myo*-inositol 4,5-cyclic phosphate (**25**)].

Up to now all isomers of inositol (poly)phosphates detected in mammalian cells have been synthesized. Apart from the synthesis of positional isomers of inositol (poly)phosphates, additional structure-activity relationships can be gathered from the synthesis of IP$_3$ analogues containing phosphate modified functions or phosphate isosteric groups (see Figure 2). Especially the latter topic has not gained much

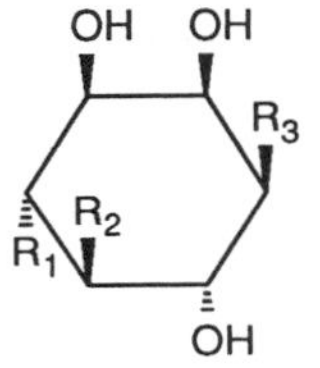

9	$R_1=R_2=OPO_3^{2-}$, $R_3=OPO_3R^-$ (**45**)
10	$R_1=R_3=OPO_3^{2-}$, $R_2=CH_2PO_3^{2-}$ (**46**)
11	$R_1=R_2=R_3=OP(H)O_2^-$ (**47**)
12	$R_1=R_2=R_3=OP(S)O_2^{2-}$ (**48**)
13	$R_1=R_3=OPO_3^{2-}$, $R_2=OP(S)O_2^{2-}$ (**49**)
14	$R_1=R_2=OPO_3^{2-}$, $R_3=OP(S)O_2R^-$ (**50**)

Figure 2. IP$_3$ analogs containing phosphate-modified functions or phosphate isosteric groups.

attention yet. The only isosteric IP$_3$ analogues that have been published so far are synthetic derivatives containing a methylene phosphonate (e.g. **10**) or phosphorthioate moieties [e.g. **12**, **13** and **14** (R=H)]. The 1-phosphatediesters (*45*) (**9**, R= sn-3-glycero, glycoaldehyde, glycolic acid and N-octyl-2-aminoethyl), 5-methylene phosphonate **10** (*46*), 1,4,5-trisphosphorothioate **12** (*48*), 5-phosphorothioate **13** (*49*) and 1-phosphorthioate **14** (*50*) (substituted, R= N-[{2-acetoxyethyl}-N-methyl]amino-7-nitro-2,1,3-benzoxadiazole) derivatives turned out to be IP$_3$ agonists in releasing intracellular Ca^{++}. No biological activity of the hydrogen phosphonate analogue **11** (*47*) has been reported yet.

It is obvious that inositol phosphate analogues which might be cell permeable would find considerable application. To this end the synthesis of 2,3,6-tributyryl *myo*-inositol 1,4,5-trisphosphate (*41*) (**6**, Figure 1) has been addressed recently. Little attention has been addressed to the synthesis of inositol derivatives containing less negatively charged or neutral groups being isosteric to phosphate. In this paper we describe the synthesis of a less negatively charged and a neutral analogue, i.e. inositol 1,4,5-trissulphate (**31**) and inositol 1,4,5-trissulphonamide (**32**) respectively (*51*), both compounds isosteric to IP$_3$.

Synthesis of D-*myo*-Inositol 1,5-bisphosphate and 3,5-bisphosphate

In order to prepare the suitably protected key intermediate, two different routes were explored, i.e the classical 'bis-ketal' route (*52*) and the 'orthoester' route (*53*). The former route leads to intermediate **15** (*54*), but turned out to be rather tedious in the next step of the synthesis since no regioselectivity could be achieved in protecting the 4-OH of the 4,5-*vic*-diol of **15** (Scheme 1). In order to circumvent the regioselective protection of the 4,5-*vic*-diol, we examined the applicability of inositol orthoformate **18** (*55*). Perbenzylation of orthoformate **18** using sodium hydride/benzyl bromide in DMF afforded the fully protected inositol **19**, which on removal of the orthoester furnished the meso 2,4,6-tri-O-benzyl derivative **20** (Scheme 2). Benzylation of triol **20** (1 mmol) under phase transfer conditions using benzyl bromide (2 eq.), tetra-n-butyl-ammonium iodide (0.25 eq.) and sodium hydroxide (3 ml, 5% solution in water) in dichloromethane (15 ml) for 16h at 45°C

led predominantly to the desired 1,2,4,6-tetra-O-benzyl *myo*-inositol **17** (*56*) (71 %
yield from **18**), the structure of which was readily deduced from its dissymmetric 2D
^{1}H-NMR spectrum [^{1}H-NMR (200 MHz), (CDCl$_3$): δ 3.35-3.50 (c, 2H,H-1, H-5);

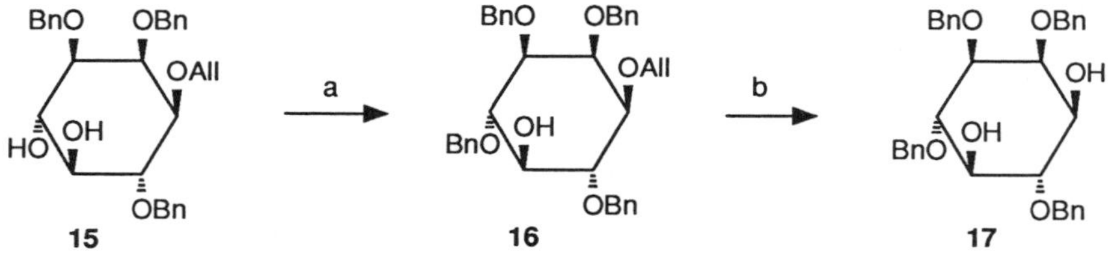

a: BaO/Ba(OH)$_2$/BnBr, 0°C, 20%; b: Ir[COD(PMePh$_2$)$_2$]PF$_6$/H$_2$ then HCl/dioxane/MeOH

Scheme 1

3.53 (dd, 1H, H-3); 3.69 (t, 1H, H-4); 3.88 (t, 1H, H-6); 4.03 (t, 1H, H-6); 4.65-5.05
(c, 8H, 4xCH$_2$); 7.20-7.40 (c, 20H, 4xPh)]. Key intermediate **17** was subjected to
optical resolution prior to phosphorylation experiments. We took into account that
the optical resolution of an inositol derivative containing an 1,2-diol was realized
through a regioselective formation of diastereoisomeric 1-O-menthoxyacetates

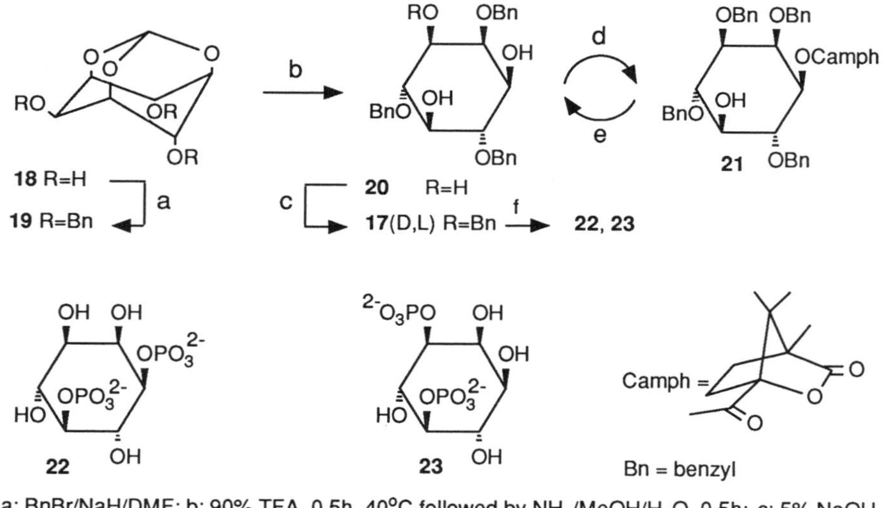

a: BnBr/NaH/DMF; b: 90% TFA, 0.5h, 40°C followed by NH$_3$/MeOH/H$_2$O, 0.5h; c: 5% NaOH
CH$_2$Cl$_2$/BnBr/Bu$_4$NI, 16h, 45°C; d: 1.1 eq. (-)camphanic acid chloride in pyridine, 0°C; e: Li-
OH in dioxane/MeOH, 1.5h, rt; f: (CNCH$_2$CH$_2$O)$_2$PN(CH$_2$CH$_3$)$_2$/1-H-tetrazole in CH$_2$Cl$_2$/
CH$_3$CN, 0°C- rt and then tert-BuOOH/Et$_3$N and 0.2N NaOH/MeOH; 10% Pd on C/H$_2$/DMF

Scheme 2

(*57-59*). After a number of experiments, we found that a diastereomeric mixture of
the 1-O-camphanates **21**, prepared in a regioselective reaction of 1,5-diol **17** with
1.1 equivalent of (-)-camphanic acid chloride in pyridine at 0°C, could be separated
efficiently by silica gel column chromatography. The optical purity of the separated

diastereoisomers (**21D**, 47% yield, R_f 0.38, CH_2Cl_2/EtOAc, 97:3, v/v and **21L**, 44 % yield, R_f 0.55, CH_2Cl_2/EtOAc, 97:3, v/v) could be checked by ^{1}H-NMR analysis since the diastereoisomeric camphanate methyl resonances exerted different chemical shifts. Alkaline hydrolysis (LiOH in 1:1 dioxane/methanol for 1.5 h at 20°C) of **21D** and **21L** afforded the enantiomerically pure diols **17D** ($[\alpha]_D$ 10.0, c=1, CHCl$_3$) and **17L** ($[\alpha]_D$ -9.1, c=1, CHCl$_3$), the absolute configurations of which were determined by comparison of the optical rotations with that of a sample synthesized in two steps from known enantiomeric **15** (*54*) (Scheme 1).

Phosphorylation of **17D** and **17L** was performed efficiently with a phosphoramidite reagent. Thus, to a stirred solution of the 1,5-diol   (0.2 mmol) and bis(2-cyanoethyl)-*N,N*-diethylphosphoramidite (*60*) (0.6 mmol) in dichloromethane was added 1 ml of a 1M solution of 1*H*-tetrazole in acetonitrile. After 1 hr, a solution of excess *tert*-butyl hydroperoxide and triethylamine in dichloromethane is added to the reaction mixture and after a further period of 4 hr, the mixture was concentrated and applied to a column of Sephadex LH-20. The appropriate fractions were treated with 0.2 N NaOH/MeOH/dioxane and neutralized (Dowex 50W, H$^+$-form). Subsequent hydrogenolysis (10% Pd/C) afforded the title compounds D-*myo*-inositol 1,5-bisphosphate **22** and 3,5-bisphosphate **23** in about 95% purity. Since samples of a higher purity are required, the individual derivatives were separately subjected to DEAE column chromatography (1x30 cm^2, 0.2-0.8 M NH$_4$OAc). The appropriate fractions [detected by spot test using modified Jungnickel's reagent (*61*)] were lyophilized to furnish highly pure **22** (-D-, 83% yield from **17D**, $[\alpha]_D$ 6.0, c=0.5, H$_2$O) and **23** (-L-, 81% yield from **17L**, $[\alpha]_D$ -5.9, c=0.5, H$_2$O). The structure and homogeneity of **22** and **23** was confirmed by ^{1}H, ^{13}C and ^{31}P NMR spectroscopy. [^{1}H NMR (360 MHz, D$_2$O): δ 3.67 (dd, H-3); 3.79 (t, H-4); 3.90 (c, H-5, H-6); 4.02 (m, H-1); 4.27 (t, H-2)].

Synthesis of *myo*-Inositol 4,5-cyclic Phosphate

Inositol 1,2-cyclic phosphate has been synthesized by reaction of a perbenzylated 1,2-diol with N-methylpyridinium phosphodichloridate (*62*) followed by catalytic hydrogenation. This cyclic inositol phosphate contains a relatively stable phosphodiester between the cis orientated hydroxyl groups at positions 1 and 2. A

a: N-methylpyridinium phosphodichloridate; b: 10% Pd on C/H$_2$/DMF   Scheme 3

similar approach has been used by us in order to prepare the less stable, trans-di-equatorial 4,5-cyclic phosphate **25**. Thus, rac. 1,2,3,4-tetra-O-benzyl *myo*-inositol **24** (*63*) was treated with N-methylpyridinium phosphodichloridate (1.0 equivalent) at 0°C for 1 hour. (Scheme 3). The reaction mixture was concentrated and applied to Sephadex LH20 column chromatography (eluent: CH_2Cl_2/MeOH, 2:1, v/v). Catalytic hydrogenation of the appropriate fractions afforded *myo*-inositol 4,5-cyclic phosphate **25** in 81% yield starting from **24**. ^{31}P NMR spectroscopy revealed the presence of one resonance at δ 17.4 ppm, a chemical shift characteristic for a cyclic phosphate five-membered ring. ^{13}C NMR spectroscopy indicated the presence of doublets at δ 71.1 ppm (C-4, $J_{C,P}$ 3 Hz) and δ 71.9 ppm (C-5, $J_{C,P}$ 3 Hz).

Synthesis of 1,4,5-trissulphated-, 1,4,5-trissulphamoylated- and 1,4,5-trisphosphorylated *myo*-Inositol

Properly protected **26** was obtained from *myo*-inositol by the procedure described by Gigg et al (*54*). Regioselective stannylene-mediated allylation (*64*) of **26** and subsequent benzylation afforded fully protected **27** (73% yield, see Scheme 4). Treatment of **27** with cyclooctadiene-bis(methyldiphenylphosphine)-iridium hexafluorophosphate (*65*) afforded **28** (100% yield), which on acid hydrolysis (dioxane/methanol/0.1 N HCl) furnished key-intermediate triol **29** (96% yield). We now turned our attention to the introduction of the different functional groups, i. e. sulphate and sulphonamide (Scheme 5). Treatment of **29** (0.2 mmol) for 16 hr at

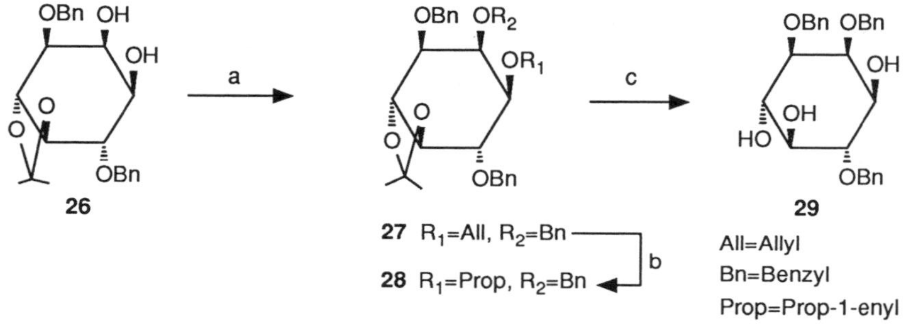

a: Bu_2SnO, Bu_4NI, AllBr/toluene; NaH, BnBr/DMF; b: $Ir[COD(PMePh_2)_2]PF_6/H_2$; c: HCl MeOH/dioxane

Scheme 4

50°C with triethylamine sulphur trioxide complex (*66*) (3 mmol) in DMF provided a protected sulphated inositol intermediate (R_f 0.35, EtOAc: Pyr: AcOH: H_2O, 11: 7: 1.6: 4). The latter was purified by Sephadex LH-20 column chromatography (eluent DMF containing 0.5% triethylamine, the appropriate fractions were treated with sodium bicarbonate (1 eq./ OSO_3^-) prior to concentration and silica gel column chromatography (eluent: CH_2Cl_2/MeOH, 4:1). Subsequent hydrogenolysis (10% Pd on C in 4:1 DMF/H_2O) furnished *myo*-inositol 1,4,5-trissulphate **31** in a yield of 86% from **29**. In order to obtain the neutral 1,4,5-trissulphamoylated derivative **32**, triol **29** (0.2 mmol) was treated with sulphamoyl chloride (*67*) (1.2 mmol) and NaH (1.2 mmol) in DMF at 0°C for 2 hr to give, after silica gel column chromatography (eluent: CH_2Cl_2/MeOH, 95:5), the protected intermediate (R_f 0.6, CH_2Cl_2/MeOH, 99:1). Subsequent debenzylation (10% Pd on C in DMF/H_2O) gave inositol 1,4,5-trissulphonamide **32** (R_f 0.25, RP18 silica, EtOH/H_2O, 3:1) in 84% yield from **29**. Starting from compound **29** we also synthesized (*rac.*) IP_3 itself (**30**), as a reference compound, applying the same procedure used for the synthesis of compounds **22** and **23**.

The identity and homogeneity of the compounds **30**, **31** and **32** were established by NMR spectroscopy and (FAB) mass spectrometry. The ^{1}H-NMR spectral data are listed in the Table 1.

Biological Evaluation and Conclusions

The preliminary biological activity of the compounds was determined in three assays: a permeabilized human blood platelet aggregation model (*68*), a radioligand binding assay using rat cerebellar membranes and a Ca^{++}-mobilization assay

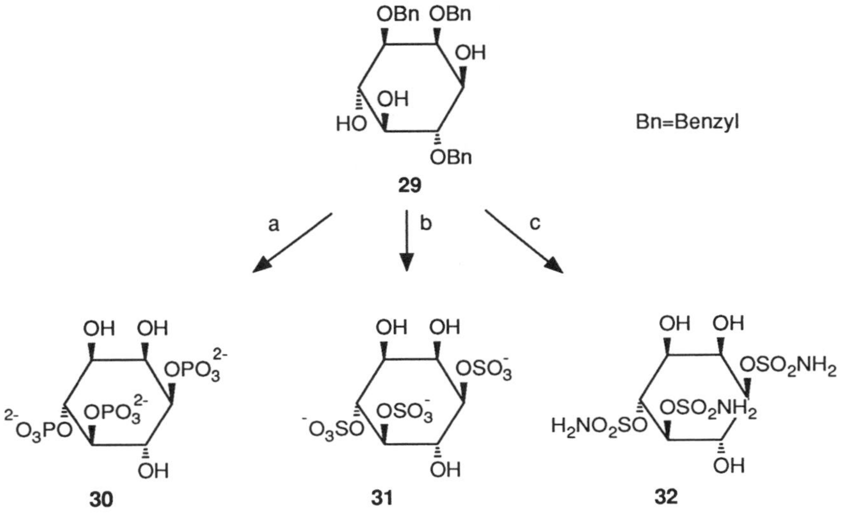

Scheme 5

a: 1. $(CNCH_2CH_2O)_2PN(CH_2CH_3)_2$/1-H-tetrazole, 2. t-BuOOH/Et$_3$N, 3. 0.2N NaOH/ dioxan/MeOH, 4. 10% Pd on C/H$_2$/DMF/H$_2$O. b: $(CH_3CH_2)_3$N.SO$_3$/DMF then 10% Pd/ H$_2$/DMF/H$_2$O. c: NH$_2$SO$_2$Cl/NaH/DMF then 10% Pd on C/H$_2$/DMF.

applying a human neuroblastoma SY5Y cell line (69). The latter two assays were performed in the laboratory of Prof. S. R. Nahorski.

D-*myo*-inositol 1,5-bisphosphate (22) and 3,5-bisphosphate (23) were tested on their affinity for the IP$_3$ binding site and in our functional bio-assay (saponin permeabilized human bloodplatelet aggregration). In neither assay did the compounds exhibit any activity. These results suggest that the two phosphate groups in D-1,5- and 3,5-bisphosphate are not sufficient for a molecular interaction with the receptor protein. Preliminary results also indicate that inositol 4,5-cyclic

Table 1 ^{1}H NMR chemical shifts (360 MHz) of compounds **30, 31** and **32**

inositol 1,4,5-		H$_1$	H$_2$	H$_3$	H$_4$	H$_5$	H$_6$
trisphosphate	**30**	4.10 c	4.26 c	3.71 dd	4.26 c	4.10 c	3.88 t
trissulphate	**31**	4.27 dd	4.41 t	3.85 dd	4.52 t	4.29 t	3.96 t
trissulphonamide	**32**	4.55 dd	4.57 t	3.97 dd	4.79 t	4.64 t	4.16 t

phosphate **25** is inactive in our aggregration assay. The results obtained support Irvine's prediction that the vicinal 4,5-bisphosphate is an essential structural requirement for IP$_3$ agonistic activity (33) on mammalian receptors.

It has previously been shown that the anticoagulant drug heparin, a sulphated glycosaminoglycan, may compete with IP$_3$ for its receptor binding (70) and for inhibiting the IP$_3$-induced Ca^{++} release (71). It should be mentioned that it is not yet clear whether the affinity of heparin for the IP$_3$ receptor can be attributed to binding of sulphate groups in regions normally accomodating the IP$_3$ phosphate groups. We reasoned that in this respect the biological evaluation of inositol 1,4,5-trissulphate **31** would be of considerable interest. However, the sulphated IP_3 analogue **31** did not display any IP_3 antagonistic or agonistic activity in our platelet aggregation model nor in a IP_3 binding assay. This finding suggest that sulphated carbohydrates, like heparin, do not compete at the IP_3 binding site itself. In order to get more information the binding affinity of two highly sulphated synthetic heparin fragments (i.e. **33** and **34**, Scheme 6) for the IP_3 binding site (on cerebellar membranes) was investigated. Compound **33** represents the well-defined pentasaccharide region of

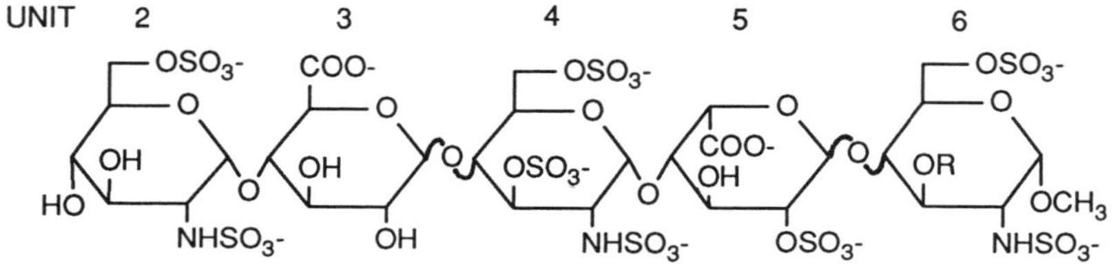

33 R=H
34 R=SO$_3^-$

Scheme 6

heparin that activates antithrombin III. Organic synthesis of **33** (72) and many analogues facilitated the definition of structure-activity relationships (73), which led to the introduction of an extra sulphate ester at unit 6 to give compound **34** (74). The latter compound, being approximately twice as active as compound **33**, is the most potent analogue found thus far to activate antithrombin III. Surprisingly, compound **33** as well as the highly sulphated compound **34** showed no affinity for the IP_3 binding site. Recently it was found that de-N-sulphated heparin lacks the IP_3 antagonistic activity found with heparin (75). Therefore it was suggested that binding of the heparin fragment to the IP_3 receptor depends on the relative positions of the different O- and N-sulphate esters on the glucosamine units (76). Compound **34** contains two glucosamine units bearing three sulphate esters (i.e. unit 4 and 6), a property not found in any other region of the heparin glycosaminoglycan. Hence, it is highly unlikely that the affinity of heparin for the IP_3 binding site can be attributed to the presence of one specific glucosamine residue containing three sulphate esters.

Initial evaluation indicate that the neutral 1,4,5-trissulphonamido derivative **32** is biologically inactive. An explanation for this may be the enormous difference in networks of hydrogen bonds, salt bridges and ionic interactions which are expected to be formed between a charged (i.e. phosphorylated or sulphated) or a neutral substrate at the IP_3 binding site.

Acknowledgements

Authors wish to thank: Prof. Dr. S. R. Nahorski (University of Leicester) for performing the radio ligand assay and the Ca^{++}-mobilizing assay; W. J. G. Joosten and T. G. van Dinther for performing the platelet aggregation assay; Dr. D. G. Meuleman (Organon, Oss) and Dr. D. Nicholson (Organon, Newhouse) for useful discussions.

Literature Cited

1. Hokin, L. E. *Annu. Rev. Biochem.* **1985**, 54, 205.
2. Nishizuka, Y. *Nature* **1984**, 308, 693.

3.   Berridge, M. J. *Sci. Am.* **1985**, 253, 124.
4.   Berridge, M. J. *J. Exp. Biol.* **1986**, 124, 323.
5.   Abdel-Latif, A. A. *Pharmacol. Rev.* **1986**, 38, 227.
6.   *Inositol Lipids in Cellular Signalling*, Michell, R. H.; Putney, J. W. Jr., Eds.; Current Communications in Molecular Biology; Cold Spring Harbor: 1987.
7.   Nahorski, S. R. *Trends Neurosci.* **1988**, 11, 444.
8.   *Inositol Lipids in Cell Signalling*, Michell, R. H.; Drummond, A. H.; Downes C. P., Eds.; Academic Press: London, 1989.
9.   Nahorski, S. R.; Potter, B. V. L. *Trends Pharmacol. Sci.* **1989**, 10, 139.
10. Joseph, S. K.; Williamson, J. R. *Arch. Biochem. Biophys.* **1989**, 237, 1.
11. Berridge, M. J.; Irvine, R. F. *Nature* **1989**, 341, 197.
12. Downes, C. P.; MacPhee, C. H. *Eur. J. Biochem.* **1990**, 193, 1.
13. Nahorski, S. R. *Br. J. Clin. Pharmac.* **1990**, 30, 23S.
14. Majarus, P. W.; Ross, T. S.; Cunningham, T. W.; Caldwell, K. K.; Jefferson, A. B.; Bansal, V. S. *Cell* **1990**, 63, 459.
15. Rana, R. S.; Hokin, L. E. *Physiological Reviews* **1990**, 70, 115.
16. Streb, H.; Irvine, R. F.; Berridge, M. J.; Schultz, I. *Nature* **1983**, 306, 67.
17. Berridge, M. J.; Irvine, R. F. *Nature* **1984**, 312, 315.
18. Rando, R. R. *Faseb J.* **1988**, 2, 2348.
19. Chilvers, E. R.; Kennedy, E. D.; Potter, B. V. L. *Drugs News and Perspectives* **1989**, 2, 342 and ref. therein.
20. Ozaki, S.; Watanabe, Y.; Ogasawara, T.; Kondo, Y.; Shiotani, N.; Nishii, H.; Matsuki, T. *Tetrahedron Lett.* **1986**, 27, 3157.
21. Gigg, J.; Gigg, R.; Payne, S.; Conant, R. *J. Chem. Soc., Perkin Trans I* **1987**, 423.
22. Billington, D. C. *Chem. Soc. Rev.* **1989**, 18, 83 and ref. therein.
23. Vacca, J. S.; deSolms, S. J.; Huff, J. R.; Billington, D. C.; Baker, R.; Kulagowski, J. J.; Mawer, I. M. *Tetrahedron* **1989**, 45, 5679 and ref. therein.
24. Potter, B. V. L. *Natural Product Reports* **1990**, 1 and ref. therein.
25. IP_3 phosphorylation by a 3-kinase yields inositol 1,3,4,5-tetrakisphosphate, another putative second messenger. See: Batty, I. R.; Nahorski, S. R.; Irvine, R. F. *Biochem. J.* **1985**, 232, 211.
26. Bourne, H. R. *Nature* **1986**, 321, 814.
27. Murphy, P. M.; Eide, B.; Goldsmith, P. et al. *FEBS Lett.* **1987**, 221, 81.
28. Harnett, M. M.; Klaus, G. G. B. *Immunology Today* **1988**, 9, 315.
29. Nishizuka, Y. *Nature* **1988**, 334, 661.
30. Rhee, S. G.; Suh, S-G.; Ryu, S-H.; Lee, S. Y. *Science* **1989**, 244, 546.
31. Beavo, J. A.; Reifsnyder, D. H. *Trends Pharmacol. Sci.* **1990**, 11, 150.
32. Worley, P. F.; Baraban, J. B.; Colvin, S. S.; Snyder, S.H. *Nature* **1987**, 325, 159.
33. Irvine, R. F.; Brown, K. D.; Berridge, M. J. *Biochem. J.* **1984**, 222, 269.
34. Willcocks, A. L.; Strupish, J.; Irvine, R. F.; Nahorski, S. R. *Biochem. J.* **1989**, 257, 297.
35. Hirata, M.; Watanabe, Y.; Ishimatsu, T.; Ikebe, T.; Kimura, Y.; Yamaguchi, K.; Ozaki, S.; Koga, T. *J. Biol. Chem.* **1989**, 264, 20303.
36. Kozikowski, A. P.; Fauq, A. H. *J. Am. Chem. Soc.* **1990**, 112, 7403.
37. Seewald, M. J.; Aksoy, I. A.; Powis, G.; Fauq, A. H.; Kozikowski, A. P. *J. Chem. Soc. Chem. Commun.* **1990**, 1638.
38. Polokoff, M. A.; Bencen, G. H.; Vacca, J. P.; de Solms, J.; Young, S. D.; Huff, J. R. *J. Biol. Chem.* **1988**, 263, 11922.
39. Ley, S. V.; Strernfeld, F. *Tetrahedron Lett.* **1988**, 29, 5305.
40. Ley, S. V.; Parra, M.; Readgrave, A. J.; Sternfeld, F.; Vidal, A. *Tetrahedron Lett.* **1989**, 30, 3557.
41. Tegge, W. *Ph. D. Thesis*, University of Bremen, **1986**. Biological evaluation: Schultz, C. *Ph. D. Thesis*, University of Bremen, **1989**.

42. Baker, R.; Leeson, P. D.; Liverton, N. J.; Kulagowski, J. J. *J. Chem. Soc., Chem. Commun.* **1990**, 462.
43. Schultz, C.; Gebauer, G.; Metschies, T.; Rensing, L.; Jastorff, B. *Biochem. Biophys. Res. Commun.* **1990**, 166, 1319.
44. Westerduin, P.; Willems, H. A. M.; van Boeckel, C. A. A. *Tetrahedron Lett.* **1990**, 31, 6915.
45. Henne, V.; Mayr, G. W.; Grabowski, B.; Koppitz, B.; Soeling, H-D. *Eur. J. Biochem.* **1988**, 174, 95.
46. Falck, J. R.; Abdali, A.; Wittenberger, S. J. *J. Chem. Soc. Chem. Commun.* **1990**, 953.
47. Dreef, C. E.; von der Marel, G. A.; van Boom, J. H. *Recl. Trav. Chim. Pays-Bas* **1987**, 106, 512.
48. Cooke, A. M.; Gigg, R.; Potter, B. V. L. *J. Chem. Soc. Chem. Commun.* **1987**, 1525.
49. Cooke, A. M.; Noble, N. J.; Payne, S.; Gigg, R; Potter, B. V.L. *J. Chem. Soc. Chem. Commun.* **1989**, 269.
50. Lampe, D.; Potter, B. V. L. *J. Chem. Soc. Chem. Commun.* **1990**, 1500.
51. Westerduin, P.; Willems, H. A. M.; van Boeckel, C. A.A. *Tetrahedron Lett.* **1990**, 31, 6919.
52. Angyal, S. J.; Tate, M. E.; Gero, S. D. *J. Chem. Soc.* **1961**, 4116. Gigg, R.; Warren, C. D. *J. Chem. Soc.(C)* **1969**, 2367.
53. Billington, D. C.; Baker, R.; Kulagowski, J. J.; Mawer, I. M.; Vacca, J. P.; deSolms, S. J.; Huff, J. R. *J. Chem. Soc. Perkin Trans I* **1989**, 1423.
54. Gigg, J.; Gigg, R.; Payne, S.; Conant, R. *J. Chem. Soc. Perkin Trans I* **1987**, 1757.
55. Lee, H. W.; Y. Kishi, Y. *J. Am. Chem. Soc.* **1985**, 107, 4402.
56. 1,2,4,6-tetra-O-benzyl-*myo*-inositol was isolated by Billington (ref. 22) as an undesired product in a synthesis towards *myo*-inositol 1,3-bisphosphate.
57. Ozaki, S.; Watanabe, Y.; Ogasawara, T.; Kondo, Y.; Shiotani, N.; Nishii, H.; Matsuki, T. *Tetrahedron Lett.* **1986**, 27, 3157.
58. Watanabe, Y.; Ogasawara, T.; Nakahira, H.; Matsuki, T.; Ozaki, S. *Tetrahedron Lett.* **1988**, 29, 5259.
59. Ozaki, S.; Kohno, M.; Nakahira, H.; Bunya, M.; Watanabe, Y. *Chemistry Lett.* **1988**, 77.
60. Reese, J. B.; Ward, J. G. *Tetrahedron Lett.* **1987**, 28, 2309.
61. Dittmer, J. C.; Lester, R. L. *J. Lipid Res.* **1964**, 5, 126; Vaskovsky, V. E.; Latyshev, N. A. *J. Chromatogr.* **1975**, 115, 246.
62. Watanabe, Y.; Ogasawara, T.; Nakahira, H.; Matsuki, T.; Ozaki, S. *Tetrahedron Lett.* **1988**, 29, 5259.
63. Hamblin, M. R.; Potter, B. V. L.; Gigg, R. *J. Chem. Soc. Chem. Commun.* **1987**, 626.
64. David, S; Hanessian, S. *Tetrahedron* **1985**, 41, 643.
65. Oltvoort, J. J.; van Boeckel, C. A. A.; de Koning, J. H.; van Boom, J. H. *Synthesis* **1981**, 305.
66. Gilbert, E. E. *Chem. Soc. Rev.* **1962**, 62, 549.
67. Sulphamoyl chloride was prepared from chlorosulphonyl isocyanate by controlled hydrolysis.
68. Israels, S. J.; Robinson, P.; Docherty, J. C.; Gerrard, J. M. *Thromb. Res.* **1985**, 40, 499. Meuleman, D. G. Akzo Pharma, Organon Scientific Development Group, unpublished;
69. Lambert, D. G.; Nahorski, S. R. *Biochem. J.* **1990**, 265, 555.
70. Worley, P. F.; Baraban, J. M.; Colvin, J. S.; Snyder, S. H. *Nature* **1987**, 325, 159.
71. Hill, T. D.; Berggren, P-O.; Boynton, A. L. *Biochem. Biophys. Res. Comm.* **1987**, 149, 897.

72. Petitou, M.; Duchaussoy, P.; Lederman, I.; Choay, J.; Jacquinet, J. C.; Sinay, P.; Torri, G. *Carbohydr. Res.* **1987**, 167, 67.

73. Petitou, M. In Heparin, Chemical and Biological Properties, Clinical Applications; Lane, D. A.; Lindahl, V.; Arnold, E., Eds.; London, **1989**, 65.

74. van Boeckel, C. A. A.; Beetz, T.; van Aelst, S. F. *Tetrahedron Lett.* **1988**, 29, 803.

75. Ghosh, T. K.; Eis, P. S.; Mullaney, J. M.; Ebert C. L.; Gill, D. L. *J. Biol. Chem.* **1988**, 263, 11075.

76. Chopra, L. C.; Twort, C. H. C.; Ward, J. P. T.; Cameron, I. R. *Biochem. Biophys. Res. Comm.* **1989**, 163, 262.

RECEIVED March 11, 1991

Chapter 9

Chemical Modifications of Inositol Trisphosphates

Tritiated, Fluorinated, and Phosphate-Tethered Analogues

Glenn D. Prestwich and James F. Marecek

Department of Chemistry, State University of New York, Stony Brook, NY 11794–3400

Biochemical studies of proteins which bind and metabolize the inositol trisphosphate isomers $Ins(1,3,4)P_3$ and $Ins(1,4,5)P_3$ have created the need for the total synthesis of selectively modified analogs. Specifically, we describe the preparation of tritium-labeled enantiomers of both regioisomers, of fluorodeoxy $Ins(1,3,4)P_3$ and $Ins(1,4,5)P_3$, and the synthesis of a P-1 tethered aminopropylphosphodiester affinity probe for $Ins(1,4,5)P_3$ receptors.

Background

Receptor-activated cleavage of phosphatidyl inositol 4,5-bisphosphate releases the second messenger D-*myo*-inositol 1,4,5-trisphosphate (*1,2*). In turn, $Ins(1,4,5)P_3$ interacts stereospecifically (*3,4*) with membrane receptors in a variety of cells to promote the release of Ca^{2+} from intracellular stores (*5*). This process is crucial in mediating numerous cellular responses to hormones, neurotransmitters, and other cellular signals (*6,7*). Progress towards understanding the molecular basis of cell signalling as a function of the interaction of individual inositol polyphosphates with cellular targets has been extremely rapid in the past five years, as evidenced by a recent review of the literature on the chemistry and biochemistry of inositol phosphates (*8*).

Tritium-Labeled Analogs of $Ins(1,3,4)P_3$ and $Ins(1,4,5)P_3$

When we entered the field of inositol polyphosphate chemistry in 1987, high specific activity (SA) tritium-labeled $Ins(1,4,5)P_3$ and $Ins(1,3,4)P_3$ were unavailable. Commercially available material with SA = 3 Ci/mmol was obtained by separation of hydrolyzed membrane phospholipids from [^{3}H]inositol incorporation. We addressed this problem by enantioselective total radiosynthesis to produce both enantiomers of [^{3}H]$Ins(1,3,4)P_3$ (*9*) and both enantiomers of [^{3}H]$Ins(1,4,5)P_3$ (*10*) with SA values of 14-16 Ci/mmol, based on the use of high SA (58-65 Ci/mmol) sodium borotritide. It is worth noting that, beginning in 1989, high SA $Ins(1,3,4,5)P_4$ and $Ins(1,4,5)P_3$ became commercially available.

0097–6156/91/0463–0122$06.00/0

The synthesis of both D-*myo* and L-*myo*-Ins(1,3,4)P$_3$ bearing tritium labels at C-1 is illustrated for the "unnatural" L-*myo* enantiomer in Figure 1 (*9*). Resolution of a suitable protected intermediate, such as the diastereomeric camphanate esters provided both desired antipodal precursors. Oxidation of the C-1 hydroxyl to the 1-ketone followed by sodium borotritide reduction provided a 3:1 ratio of epimeric alcohols favoring the desired equatorial product. After separation by silica gel chromatography and hydrolysis of the ketal, the triol trianion was phosphorylated with excess tetrabenzyl pyrophosphate. Finally, the perbenzylated species was deprotected by hydrogenolysis to give [^{3}H]Ins(1,3,4)P$_3$.

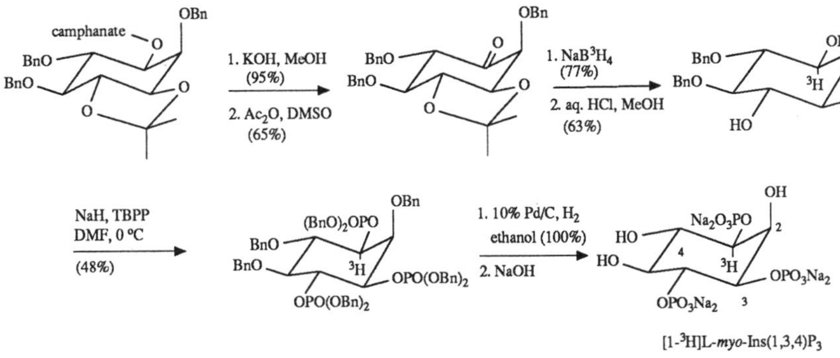

Figure 1. Synthesis of the L-*myo* enantiomer of [^{3}H]Ins(1,3,4)P$_3$.

Under conditions in which rat cerebellar membrane proteins exhibited specific binding of [^{3}H]Ins(1,4,5)P$_3$ and [^{3}H]Ins(1,3,4,5)P$_4$, neither of the [^{3}H]Ins(1,3,4)P$_3$ enantiomers showed any specific high affinity binding (Supattapone, S. and Snyder, S. H., unpublished results). However, the two enantiomers were processed differently by thrombin-stimulated human platelets (*11*). Thus, D-[^{3}H]Ins(1,3,4)P$_3$ added to saponin-permeablized platelets is hydrolyzed to an IP$_2$ species and phosphorylated to Ins(1,3,4,6)P$_4$ with relative velocities of ca. 9:1. In contrast, the unnatural L-[^{3}H]Ins(1,3,4)P$_3$ is dephosphorylated but not phosphorylated.

Using an analogous synthetic route, we used a suitably protected, resolved inositol derivative to prepare both enantiomers of [^{3}H]Ins(1,4,5)P$_3$ (*10*). In the synthesis of the natural D-*myo* enantiomer (Figure 2), we used ca. 150 mCi of 67 Ci/mmol sodium borotritide to obtain a 3:1 ratio of the desired equatorial to undesired axial alcohols. The [1-^{3}H] inositol derivative thus obtained showed a single tritium resonance appropriate for an axial triton with $^3J_{HT}$ = 2.5 Hz (to H-2$_{eq}$) and 9.8 Hz (to H-6$_{ax}$). Hydrolysis, phosphorylation, and debenzylation gave, after ion exchange chromatography, ca. 30 mCi of the [^{3}H]Ins(1,4,5)P$_3$ product.

We have observed that radiodecomposition of the high SA [^{3}H]Ins(1,4,5)P$_3$ is a serious problem at high concentrations (>10 mCi/mL). Thus, all radiolabeled materials were diluted to 0.5 mCi/mL for storage. Even this awkwardly large volume of labeled material was tenfold more concentrated than desired, based on values calculated from commercially-supplied [^{3}H]Ins(1,4,5)P$_3$. Radiochemical stability was thus greatly improved and material from this 1988 synthesis is still useful nearly two years later.

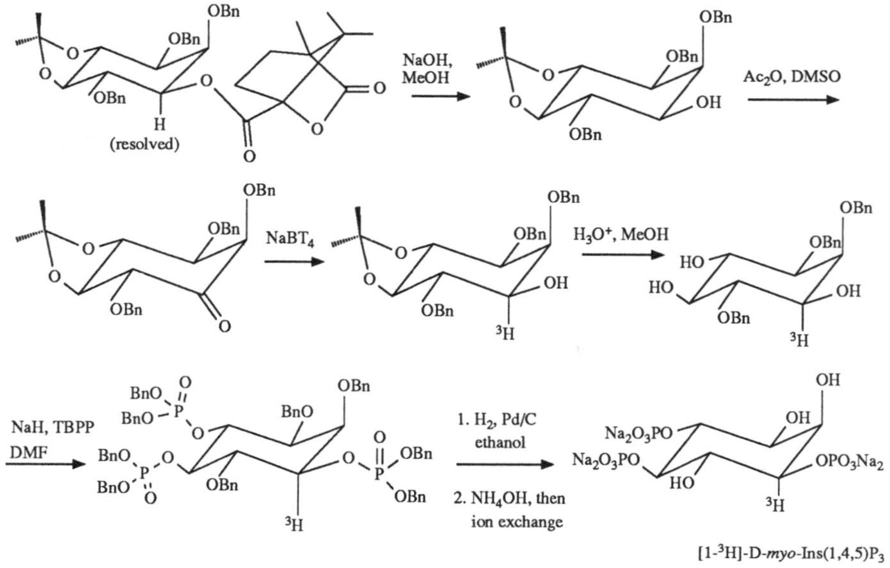

Figure 2. Asymmetric synthesis of D-*myo*-[^{3}H]Ins(1,4,5)P$_3$.

Fluorinated Analogs of Ins(1,3,4)P$_3$ and Ins(1,4,5)P$_3$

The synthetic intermediates used in synthesis of the tritium-labeled IP$_3$ enantiomers were also employed to prepare fluorodeoxy analogs of each IP$_3$ isomer. Our goal was the preparation of an antagonist or agonist with altered receptor binding properties and increased metabolic stability. Fluorodeoxy sugars have a C-F bond replacing a C-OH bond. While bond lengths and polarizations are analogous, the C-F bond can only accept but not donate hydrogen H-bond (*12*). Such compounds are potentially useful as probes for studies of the active sites of enzymes and for membrane transport studies. When we began our efforts at preparing fluorodeoxy IP$_3$ analogs, several fluorodeoxyinositols had been synthesized and studied *in vitro* (*13,14,15*), and a 2-deoxy-2-fluoro-1-phosphatidyl-*scyllo*-inositol had been prepared (*16*). Subsequently, other investigators have described the synthesis of 3-deoxy-3-fluoro inositol (*17*) and its corresponding 1,4,5-trisphosphate derivative (*18*).

We prepared the first 2-fluoro-2-deoxy Ins(1,3,4)P$_3$ analogs (*9*) as shown in Figure 3. In general, we have found that diethylaminosulfur trifluoride (DAST) fluorination of the hindered secondary alcohols in protected inositol derivatives can be very sluggish, and often proceeds with the production of two epimeric fluoro compounds. Similarly, the DAST reaction on the ketones proceeds very slowly, even at elevated temperatures. No interesting biochemical activities (e.g., kinase or phosphatase inhibition) were found for the fluorodeoxy Ins(1,3,4)P$_3$ analogs.

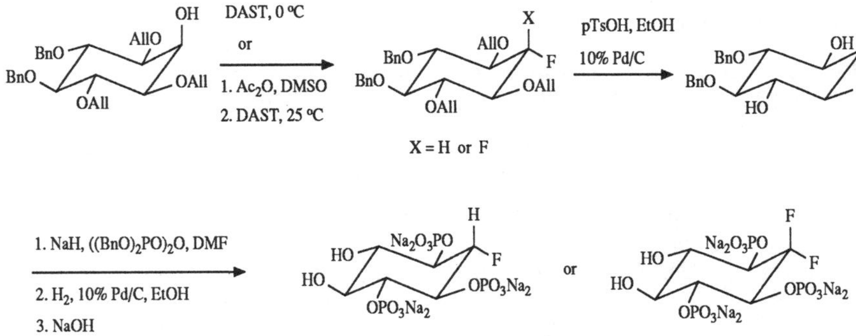

Figure 3. Synthesis of racemic 2-fluoro-2-deoxy analogs of Ins(1,3,4)$_3$.

Synthesis of the 2-fluoro-2-deoxy and 2,2-difluoro-2-deoxy Ins(1,4,5)P$_3$ analogs (19) is illustrated in Figure 4. With our scheme, we observed that both C-2 axial (*scyllo*) and C-2-equatorial (*myo*) protected inositol derivatives were converted to the _same_ 2-deoxy-2-fluoro-*scyllo* inositol derivative. Although unusual, fluorination with retention of configuration when using DAST has been observed in a number of instances. This unfortunate occurrence precluded our obtaining the desired C-2 *myo* fluoro epimer. Nonetheless, in contrast to the uninteresting biochemistry of the fluorodeoxy Ins(1,3,4)P$_3$ analogs, both the 2-fluoro-2-deoxy and 2,2-difluoro-2-deoxy-Ins(1,4,5)P$_3$ showed high affinity for the purified rat brain IP$_3$ receptor, and both activated calcium release from permeabilized cells *in vitro* (Mourey, R. J., Snyder, S. H., Marecek, J. F. and Prestwich, G. D., unpublished results).

We also observed unusual stability problems with the 2-fluoro-2-deoxy Ins(1,4,5)P$_3$. This compound underwent slow defluorination on prolonged storage (>1 month) at room temperature. Fluoride was produced (^{19}F NMR resonance at -124 ppm), and a mixture of IP$_3$ isomers resulted. The 2,2-difluoro analog and the methyl ether of the monofluoro analog were both stable under these conditions. Defluorination only occurred at pH >12, not at pH 8. By following the reaction in a sealed NMR tube using ^{19}F NMR spectroscopy, the half-life of the 2-fluoro-2-deoxy Ins(1,4,5)P$_3$ at pH 13 was estimated to be two weeks at 50 °C.

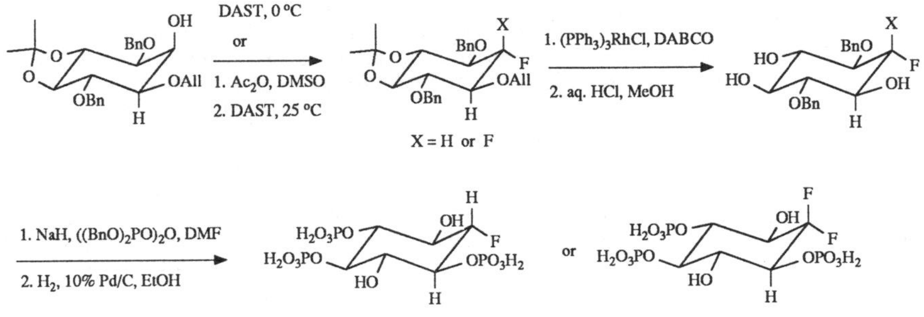

Figure 4. Synthesis of racemic 2-fluorodeoxy analogs of Ins(1,4,5)P$_3$.

Tethered and Reactive Analogs of Ins(1,4,5)P₃

Previous work. Reactive analogs and resin-immobilized analogs of Ins(1,4,5)P$_3$ are now being developed for protein characterization. A crude isomeric mixture of C-2 azidobenzoates of IP$_3$ was first described in 1985 (20). This work was then expanded to include chemically uncharacterized [^{125}I]-labeled azidosalicylamide (ASA)-β-alanyl derivatives of IP$_3$ as photoaffinity labels (21). Hirata, Watanabe, Ozaki, and their co-workers then described rational syntheses and biochemical affinities of several Ins(1,4,5)P$_3$ photo-affinity labels (22), as well as the development of two immobilized Ins(1,4,5)P$_3$ derivatives for affinity chromatography (23). Two of these materials are illustrated in Figure 5; each of these compounds is based on derivatization of the C-2 hydroxyl group of IP$_3$. Most recently, several 1-C-acyl-1,2,3,6-tetradeoxyinositol analogs have been prepared (24) but not yet tested as photoaffinity labels or as immobilized supports (Figure 5).

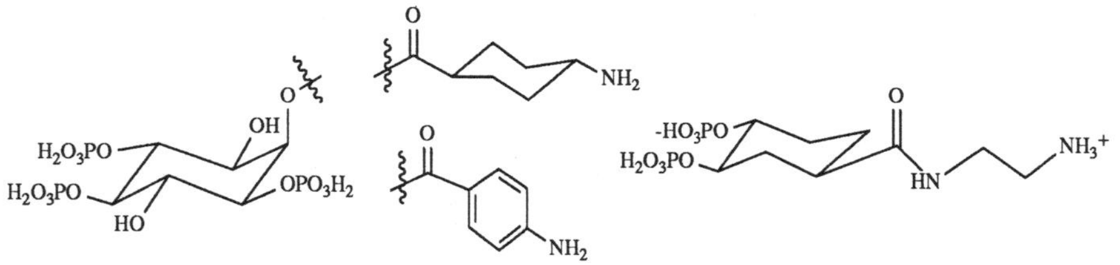

Figure 5. C-2 esters of IP$_3$ and a tetradeoxy 1-C-acyl analog of IP$_3$.

As shown below, we recently reported the first synthesis of a C-1 phosphodiester derivative of Ins(1,4,5)P$_3$ bearing a reactive aminopropyl tether. This compound was employed to make a photoaffinity label and an immobilized IP$_3$ affinity resin (25), as described below. This position for the tether appears to provide us with superior and versatile affinity reagents which retain high affinity for the receptor binding sites. Our selection of the P-1 phosphate for modification is based on two precedents. First, a series of semisynthetic IP$_3$ analogs were described (26) which stimulated calcium release from permeabilized guinea pig cells (Figure 6). Second, caged Ins(1,4,5)P$_3$ compounds, i.e., the isomeric 1-(2-nitrophenyl)ethyl phosphate esters, have been shown to possess *in vitro* receptor binding and calcium-releasing activities following laser flash photolysis (27). The C-1 phosphodiester is produced preferentially when IP$_3$ is allowed to react with the nitroaryl-diazoethane precursor (28). This P-1 caged compound, which can be employed as an efficient source of Ins(1,4,5)P$_3$, also showed intrinsic (dark) activity in stimulating Ca^{2+} release from smooth muscle sarcoplasmic reticulum and was a substrate for the 5-phosphatase.

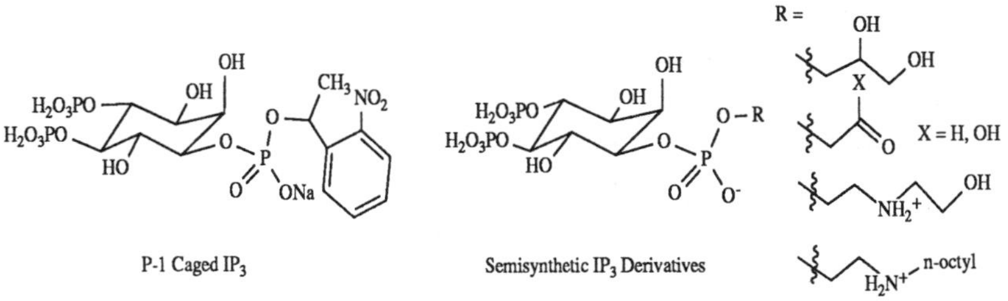

Figure 6. Caged IP$_3$ and semisynthetic IP$_3$ analogs.

Synthesis and Biochemistry of Tethered Ins(1,4,5)P$_3$ *(25)*. A phosphodiester analog of the second messenger Ins(1,4,5)P$_3$ has been synthesized and used to prepare a novel photoaffinity label and a selective bioaffinity matrix. This strategy is depicted in Figure 7.

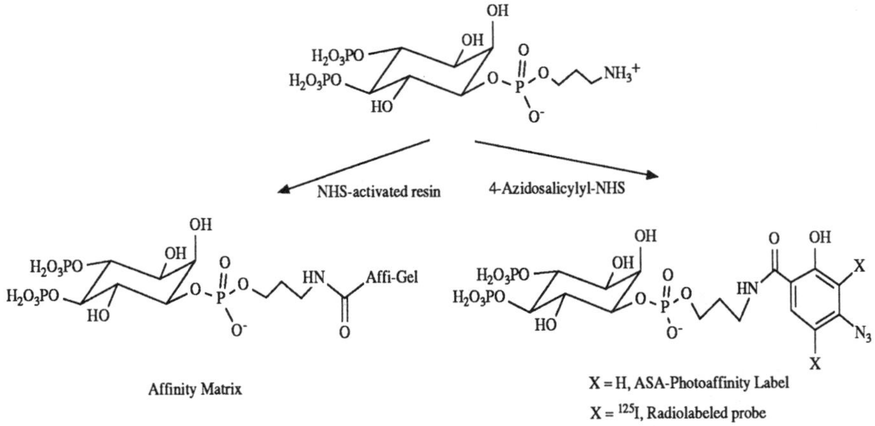

Figure 7. Design of tethered Ins(1,4,5)P$_3$ and associated receptor-targeted probes.

Using an activated phosphite bearing a Cbz-protected aminopropyl group, a selectively-protected inositol precursor was converted to an *N*-Cbz-1-*O*-(3-aminopropyl-1-phospho)-DL-*myo*-inositol in high yield following oxidation to the phosphate (Figure 8). After deprotection, followed by 4,5-bisphosphitylation and *m*CPBA oxidation, the fully benzylated 1-*O*-(3-aminopropyl) ester of Ins(1,4,5)P$_3$ was obtained in 57% yield. Hydrogenolysis (3 atm. H$_2$, 10% Pd/C, ethanol), followed by ion-exchange chromatography (Chelex, sodium form), afforded the desired tethered Ins(1,4,5)P$_3$ in 75% yield.

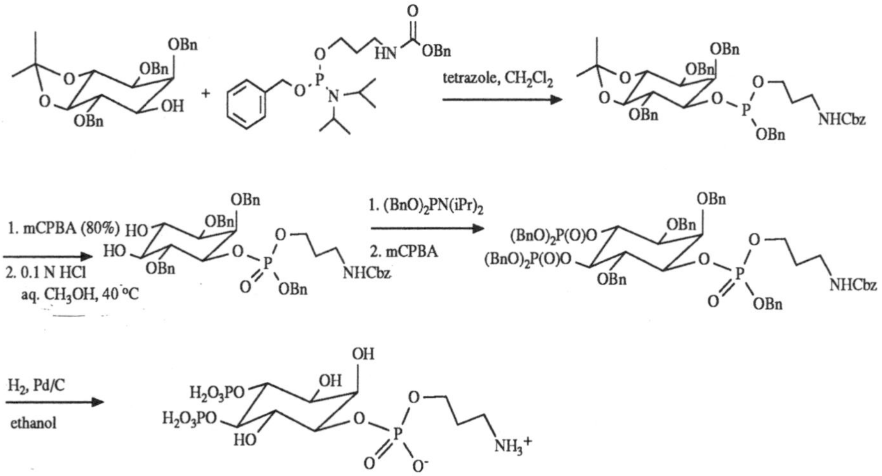

Figure 8. Synthesis of the 1-*O*-(3-aminopropyl) ester of Ins(1,4,5)P$_3$.

Reaction of the free amine with *N*-hydroxysuccinimido-4-azidosalicylate in aqueous DMF gave the 4-azidosalicylamide (ASA) derivative (*25*); this product combines photolability and a radioiodination site in the same moiety. Recently, we have radioiodinated this novel probe and successfully employed the [^{125}I]ASA derivative to photoaffinity label the partially purified rat brain IP$_3$ receptor with high selectivity and efficiency (Prestwich, G. D., Marecek, J. F., Estevez, V.A., Mourey, R. J., and Snyder, S. H., unpublished results).

Both the P-1 aminopropyl Ins(1,4,5)P$_3$ and the ASA derivative displaced [^{3}H]-D-*myo*-Ins(1,4,5)P$_3$ bound to purified rat brain Ins(1,4,5)P$_3$ receptors (*29*) with binding affinities 20-30 fold lower than that of Ins(1,4,5)P$_3$. Moreover, both analogs activated release of Ca^{2+} from liposome-reconstituted receptor preparations (*30*) at con-centrations eightfold higher than that required for Ins(1,4,5)P$_3$. The two compounds derived from tethered IP$_3$ were neither substrates nor inhibitors for the InsP$_3$ 5-phosphatase or the 3-kinase activities (*31*).

The aminopropyl tethered Ins(1,4,5)P$_3$ was immobilized on a solid support by reaction with a suspension of a resin-immobilized *N*-hydroxysuccinimide ester (Affi-Gel 10) in aq. sodium bicarbonate buffer (*24*). This resin-immobilized Ins(1,4,5)P$_3$ was employed for an efficient purification of IP$_3$ receptors from rat brain (Table 1). The phytic acid eluate from the affinity column shows a single protein band by sodium dodecyl sulfate-polyacrylamide gel electrophoresis followed by silver staining, indicating a homogeneous preparation. The aminopropyl IP$_3$ resin thus provides purified IP$_3$ receptor with a slightly greater specific activity than can be obtained by other techniques (*23,32*). While the capacity of the IP$_3$ resin is limited, small quantities of highly purified receptor can be rapidly obtained with this material.

Table 1. Purification of rat cerebellar IP_3 receptor using immobilized Ins(1,4,5)P_3. The heparin-agarose column eluate fraction was divided into two equal portions; half was applied to Con A-Sepharose, and half was applied to the IP_3 affinity resin. (Taken from Ref. 25.)

Fraction	Specific Activity (pmol/mg)	Fold Purifi- cation	% Yield
Washed membranes	0.30	1	100
1% Triton X-100 Solubilized	0.42	1.4	96
Heparin-Agarose	24	80	67
Aminopropyl IP3 Affinity Resin	230	767	29
[Concanavalin A-Sepharose]	[209]	[697]	[35]

Synthesis and Biochemistry of Other P-1 Tethered Inositol Polyphosphates. Access to the 2-O-(3-aminopropyl) ester of IP_6 was attained via phosphitylation of a suitable pentakis(allyl) ester of IP_6 and peracid oxidation to the phosphodiester (Figure 9). After removal of the allyl groups by isomerization-hydrolysis, the five free hydroxyl groups were phosphitylated and then oxidized to give the dodeca-benzyl protected hexakisphosphate. Hydrogenolysis removed all the benzyl groups to give, after ion exchange, the pentakisphosphate phosphodiester (Marecek, J. F., unpublished results). Results from A. B. Theibert and S. H. Snyder indicate that the aminopropyl IP_6 has 10% of the binding affinity (ca. 500 nM) of IP_6 itself for crude IP_6 receptor preparation. A similar reaction sequence has now provided the 2-O-(6-aminohexyl) ester of IP_6 (Marecek, J. F., unpublished results).

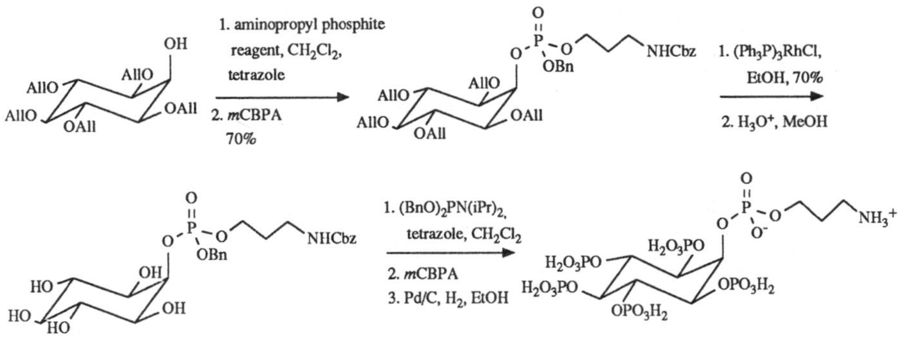

Figure 9. Synthesis of 2-O-(3-aminopropyl) ester of IP_6.

Recent progress includes synthesis of a P-1 tethered aminopropyl derivative of Ins(1,3,4,5)P_4 (Estevez, V. A. and Prestwich, G. D., submitted for publication). This derivative is now being employed in affinity purification and affinity labeling protocols for characterization of the Ins(1,3,4,5)P_4 receptor (Theibert, A. B. and Snyder, S. H., unpublished results). The successful use of the immobilized Ins(1,3,4,5)P_4 will be reported in full detail elsewhere.

Acknowledgments

We thank the Center for Biotechnology and the New York State Foundation for Science and Technology for financial support. Stimulating discussions with and unpublished results from S. Supattapone, R. J. Mourey, A. B. Theibert and others in the S. H. Snyder research group (Department of Neuroscience, Johns Hopkins University School of Medicine) are gratefully acknowledged.

Literature Cited

1. Berridge, M. J.; Irvine, R. F. *Nature* **1984**, *312*, 315-321.
2. Berridge, M. J. *Annu. Rev. Biochem.* **1987**, *56*, 159-193.
3. Willcocks, A. L.; Cooke, A. M.; Potter, B. V. L.; Nahorski, S. R. *Biochem. Biophys. Res. Commun.* **1987**, *146*, 1071-1078.
4. Strupish, J.; Cooke, A. M.; Potter, B. V. L.; Gigg, R.; Nahorski, S. R. *Biochem. J.* **1988**, *253*, 901-905.
5. Stauderman, K. A.; Harris, G. D.; Lovenberg, W. *Biochem. J.* **1988**, *255*, 677-683.
6. Nahorski, S. R. *Trends Neur. Sci.* **1988**, *11*, 444-448.
7. Berridge, M. J.; Irvine, R. F. *Nature* **1989**, *341*, 197-203.
8. Potter, B. V. L. *Nat. Prod. Rep.* **1990**, 1-25.
9. Boehm, M. F.; Prestwich, G. D. *Tetrahedron Lett.* **1988**, *29*, 5217-5220.
10. Marecek, J. F.; Prestwich, G. D. J. Labelled Compd. Radiopharm. 1989, 27, 917-925.
11. King, W. G.; Downes, C. P.; Prestwich, G. D.; Rittenhouse, S. E. *Biochem. J.* **1990**, *270*, 125-131.
12. Card, P. J. J. *Carbohydr. Chem.* **1985**, *4*, 451-487.
13. Moyer, J. D.; Reizes, O.; Malinowski, N.; Jiang, C.; Baker, D. C. ACS Symposium Series **1988**, *374*, 42-58.
14. Jiang, C.; Moyer, J. D.; Baker, D. C. *J. Carbohydr. Chem.* **1987**, *6*, 319-355.
15. Yang, S. S.; Beattie, T. R.; Shen, T. Y. *Synthetic Commun.* **1986**, *16*, 131-138.
16. Yang, S. S.; Beattie, T. R.; Shen, T. Y. *Tetrahedron Lett.* **1982**, *23*, 5517-5520.
17. Kozikowski, A. P.; Fauq, A. H.; Powis, G.; Melder, D. C. *J. Am. Chem. Soc.* **1990**, *112*, 4528-4531.
18. Kozikowski, A. P.; Fauq, A. H.; Askoy, I. A,; Seewald, M. J.; Powis, G. *J. Am. Chem. Soc.* **1990**, *112*, 7403-7404.
19. Marecek, J. F.; Prestwich, G. D.; *Tetrahedron Lett.* **1989**, *30*, 5401-5404.
20. Hirata, M.; Sasaguri, T.; Hamachi, T.; Hashimoto, T.; Kukita, M.; Koga, T. *Nature* **1985**, *317*, 723-725.
21. Ishimatsu, T.; Kimura, Y.; Ikebe, T.; Yamaguchi, K.; Koga, T.; Hirata, M. *Biochem. Biophys. Res. Commun.* **1988**, *155*, 1173-1180.
22. Hirata, M.; Watanabe, Y.; Ishimatsu, Y.; Ikebe, T.; Kimura, Y.; Yamaguchi, K.; Ozaki, S.; Koga, T. *J. Biol. Chem.* **1989**, *264*, 20303-20308.
23. Hirata, M.; Watanabe, Y.; Ishimatsu, Y.; Yanaga, F.; Koga, T.; Ozaki, S. *Biochem. Biophys. Res. Commun.* **1990**, *168*, 379-386.
24. Jina, A. N.; Ralph, J.; Ballou, C. E. *Biochemistry* **1990**, *29*, 5203-5209.
25. Prestwich, G. D.; Marecek, J. F.; Mourey, R. J.; Theibert, A. B.; Ferris, C. D.; Danoff, S. K.; Snyder, S. H. *J. Am. Chem. Soc.,* in press (1990).
26. Henne, V.; Mayr, G. W.; Grabowsi, B.; Koppitz, B.; Soling, H.-D. *Eur. J. Biochem.* **1988**, *174*, 95-101.
27. Walker, J. W.; Somlyo, A. V.; Goldman, Y. E.; Somlyo, A. P.; Trentham, D. R. *Nature* **1987**, *327*, 249-252.

28. Walker, J. W.; Feeney, J.; Trentham, D. R. *Biochemistry* **1989**, *28*, 3272-3280.
29. Worley, P. F.; Baraban, J. M.; Supattapone, S.; Wilson, V. S.; Snyder, S. H. *J. Biol. Chem.* **1987**, *262*, 12132-12136.
30. Ferris, C. D.; Huganir, R. L.; Supattapone, S.; Snyder, S. H.; *Nature* **1989**, *342*, 87-89.
31. Theibert, A. B.; Supattapone, S.; Ferris, C. D.; Danoff, S. K.; Evans, R. K.; Snyder S. H.; *Biochem. J.* **1990**, *267*, 441-445.
32. Supattapone, S.; Worley, P. F.; Baraban, J. M.; Snyder, S. H. *J. Biol. Chem.* **1988**, *263*, 1530-1534.

RECEIVED February 11, 1991

Chapter 10

Novel Routes to Inositol Phosphates Using *Pseudomonas putida* Oxidation of Arenes

Stephen V. Ley, Alison J. Redgrave, and Lam Lung Yeung

Department of Chemistry, Imperial College of Science, Technology, and Medicine, London SW7 2AY, England

Pseudomonas putida oxidation of benzene affords *cis*-3,5-cyclohexadiene-1,2-diol (**1**) which serves as a novel precursor for the synthesis of several natural products including (+)-pinitol, (+)-conduritol F, D-(-)-*myo*-inositol 1,4,5-trisphosphate and *myo*-inositol 1-phosphate. The versatility of this approach is further demonstrated by the preparation of other functionalised cyclitol derivatives, in particular 6-deoxy, 6-deoxy-6-fluoro, 6-deoxy-6-methyl and 6-methyl *myo*-inositol 1,4,5-trisphosphates.

Following the discovery and characterisation of phospholipids from the brain by Ballou in 1961 (*1*), the hypothesis by Michell in 1975 (*2*) that the receptor-controlled hydrolysis of phosphoinositides could be directly linked to cellular calcium mobilisation, and the observation by Berridge in 1984 (*3*) that D-(-)-*myo*-inositol 1,4,5-trisphosphate (IP$_3$) acts as a second messenger, a fundamental cell-signal transduction mechanism has been elucidated (*4*). This has led to a dramatic increase in interest in the inositol phosphates (*5-6*), in particular IP$_3$ which acts as a second messenger by binding to specific receptors on the endoplasmic reticulum thus stimulating the release of calcium ions from intracellular stores. The second messengers generated in the phosphatidylinositol (PI) cycle, by receptor controlled hydrolysis of phosphatidylinositol 4,5-bisphosphate (PIP$_2$), are known to regulate a large array of cellular processes including secretion, metabolism, contraction and proliferation.

Although knowledge of this phosphoinositide cell signalling system continues to grow rapidly (7) many factors remain unclear, hence there is an increased demand for improved supplies of natural products and of novel analogues to probe these mechanisms in more detail. The role of the synthetic chemist is therefore crucial to these studies.

Many of the compounds involved in the PI cycle have already been synthesised from *myo*-inositol *via* multistep protection-deprotection sequences. However, we chose to adopt a conceptually different approach which introduces the required stereogenic centres in a sequential fashion starting from benzene. This versatile strategy gives access to a wide range of derivatives not readily obtainable by conventional methods. Conversion of benzene ino *cis*-3,5-cyclohexadiene-1,2-diol (1) (8), a superb cyclitol precursor (9-15), is achieved by microbial oxidation using *Pseudomonas putida*. The significance of this biotransformation stems from the fact that an aromatic compound is converted directly into an oxidised intermediate not presently accessible by a single chemical reaction (16). Furthermore, by using substituted arenes, optically active diols can be obtained (eq.1) (10,17-20). These novel diols serve as excellent building blocks for organic synthesis.

(1) R=H

Our initial studies in this area of microbial oxidations showed how diol (1) could be readily transformed in just five steps and excellent overall yield into the cyclitol natural product (+)-pinitol (Scheme 1) (10). (+)-Pinitol has been shown to be a feeding stimulant for the larvae of the yellow butterfly *Eurema hecabe mandarina* (21) and also a larval growth inhibitor of *Heliothis zea* on soybeans (22-23). More interestingly, however, this cyclitol has recently been shown to have significant hypoglycemic and antidiabetic activity in normal and alloxan-induced diabetic albino mice and appears to be free from acute toxicity (24). During this work the unnatural antipode, (-)-pinitol, was also prepared. This was useful for comparative biological studies.

Our main efforts, however, have been directed towards the use of *cis*-3,5-cyclohexadiene-1,2-diol (1) as a precursor for the synthesis of D-*myo*-inositol 1,4,5-trisphosphate (IP₃) and related derivatives. We envisaged that the *cis*-diol functionality at C2-C3 could be carried through the synthetic route from the diene-diol (1). An epoxide opening protocol was considered to be amenable to constructing the C4-C5 *trans*-oxygenation pattern, and a second epoxide opening reaction was chosen to effect C1-C6 *trans*-oxygenation. Studies towards racemic IP₃ (Scheme 2) (11) showed that this novel approach was feasible, however we were keen to obtain optically pure materials, and especially to synthesise the unnatural series of IP₃ for biological evaluation. For this reason we chose a resolution procedure, but one which did not add any further steps to our reaction sequence.

The required cyclic epoxycarbonate (2) was prepared from (1) by treatment with sodium methoxide and dimethyl carbonate at room temperature, followed by stereoselective epoxidation using *m*-chloroperoxybenzoic acid (Scheme 3). The key step to this synthesis was the regiospecific ring-opening of (2) with an enantiomerically pure benzyl alcohol, (R)-(+)-*sec*-phenethyl alcohol (3). Reaction of (2) with (3) using a catalytic amount of HBF₄•OEt₂ in dichloromethane gave the readily separable diastereoisomeric alcohols (4) and (5) in the expected 1:1 ratio in 67% yield. After separation, benzylation of the less polar alcohol with benzyl

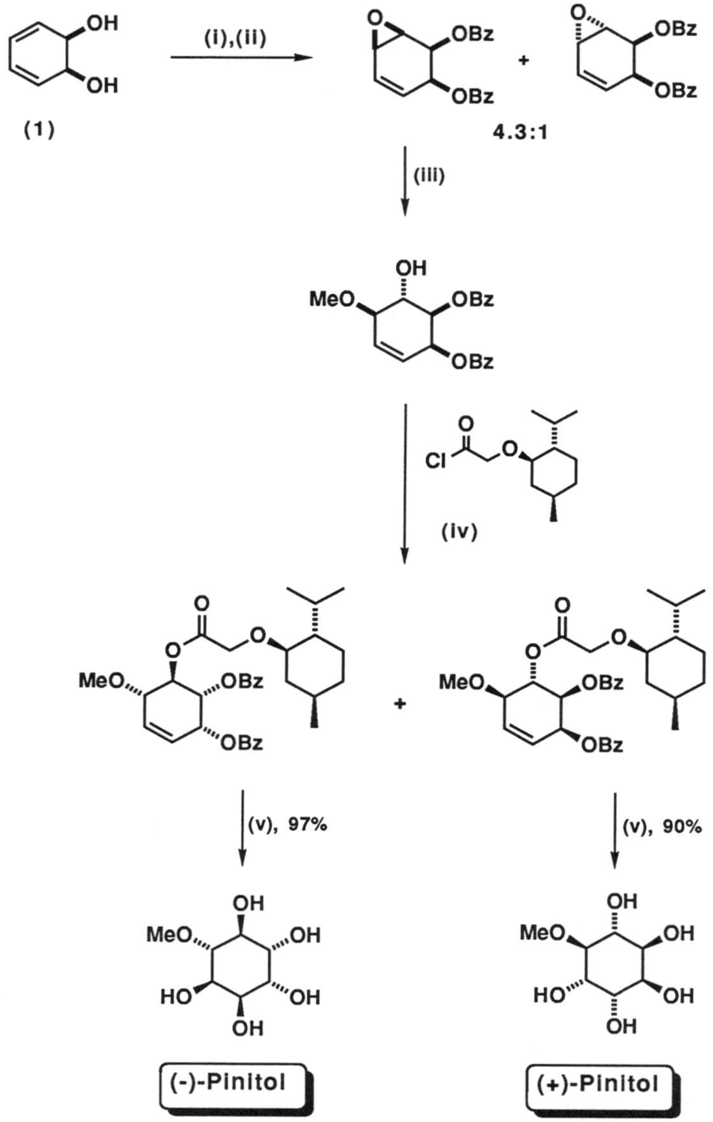

Scheme 1

(i) 2.2 BzCl, py, DMAP; (ii) 1.1 *m*CPBA, pH 8, DCE, 73%; (iii) MeOH, CSA, 100%;
(iv) DMAP, py, 87%; (v) a. OsO$_4$, NMO, ^{t}BuOH/THF/H$_2$O; b. Et$_3$N/MeOH/H$_2$O.

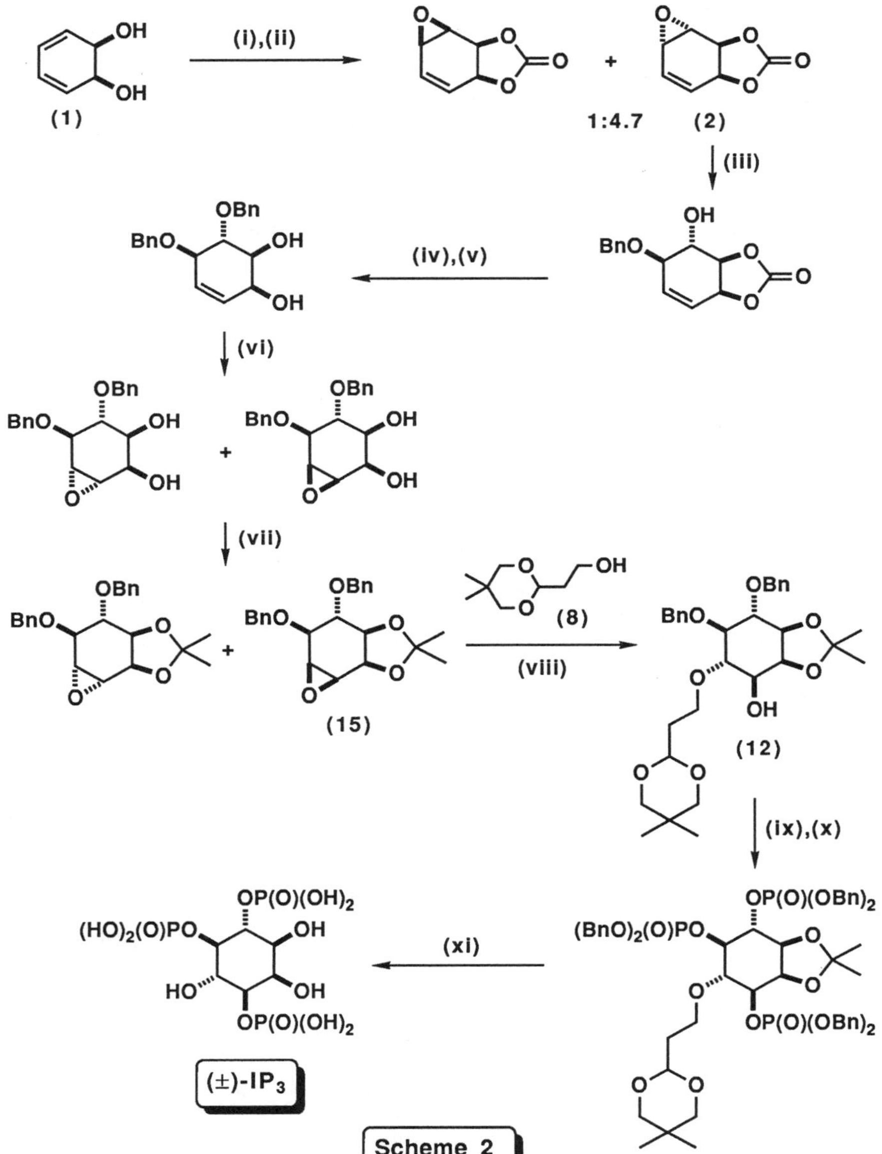

Scheme 2

(i) (MeO)₂CO, MeO⁻Na⁺, MeOH; (ii) mCPBA, DCM, 47% over 2 steps; (iii) BnOH, CSA, DCM, 85%; (iv) BnBr, Ag₂O, DMF, 81%; (v) Et₃N/MeOH/ H₂O, 100%; (vi) mCPBA, pH 8, DCM, 97%; (vii) 2,2-Dimethoxypropane, CSA, DCM, 74%; (viii) NaH, HMPA/THF, 95°C, 44h, 43%; (ix) H₂, 10% Pd-C, EtOH, 23.5h, 100%; (x) ⁿBuLi, ⁱPr₂NH, THF, tetrabenzylpyrophosphate, -30°C to RT, 62%; (xi) a. H₂, 10% Pd-C, EtOH, 4 days; b. 80% aq. TFA, 4h, 86% overall.

bromide and silver (I) oxide afforded the dibenzyl ether in excellent yield. In this way, four of the necessary oxygen functionalities were introduced with the C4,C5 hydroxyls similarly protected for later phosphorylation. Hydrolysis of the carbonate with aqueous triethylamine in methanol gave the diol (**6**) in which the hydroxyl groups were ideally placed for directed epoxidation. This epoxidation with *m*-chloroperoxybenzoic acid gave the β-epoxide in 87% yield, together with a small amount (5%) of the corresponding α-isomer which was readily removed by silica gel chromatography. The *cis*-diol was protected as the acetonide (**7**) under normal conditions.

Molecular mechanics calculations (*25*) suggested that the preferred conformation of the epoxide (**7**) was boat-like; X-ray crystallographic studies confirmed this and also the absolute stereochemistry of the system. Reaction of epoxide (**7**) with our previously developed hydroxide equivalent (**8**) (*11*) gave (**9**) and (**10**) in 37% and 58% yields, respectively. Nucleophilic ring-opening of epoxide (**15**) with methoxide was studied using molecular mechanics calculations (Figure 1). Ring-opening at C6 proceeds to give a diaxial product in a boat-like conformation (**A**) in contrast to the ring-opening at C1 which gives the diaxial product in a chair-like conformation (**C**). Calculations indicated that the boat-like conformation (**A**) was significantly lower in energy than that of the chair-like conformer (**C**). Furthermore, the initially formed diaxial boat (**A**) may be transformed to a chair conformation (**B**) which is even lower in energy.

Catalytic hydrogenolysis of (**10**) to simultaneously remove the benzyl and chiral phenethyl substituents gave the triol which, upon reaction with three equivalents of *n*-butyllithium and tetrabenzylpyrophosphate (*26-27*), produced the fully protected 1,4,5-trisphosphate (**11**) in good overall yield. Deprotection of (**11**) was readily achieved by hydrogenolysis and subsequent treatment with moist trifluoroacetic acid to give D-(-)-IP$_3$ (Scheme 3).

Preparation of the enantiomeric L-(+)-IP$_3$ followed essentially the same route as described above (*13*).

Intermediates prepared during the synthesis of (±)-IP$_3$ were also suitable precursors for cyclitol synthesis. For example, reaction of (**12**) with one equivalent of *n*-butyllithium and tetrabenzylpyrophosphate gave the 1-phosphate derivative (**13**) which, upon debenzylation by catalytic hydrogenolysis, afforded (**14**). Deprotection under acidic conditions (4:1 TFA/H$_2$O) furnished *myo*-inositol 1-phosphate, a metabolic product of the PI cycle (Scheme 4).

In an effort to extend this chemistry to the preparation of a range of novel 6-substituted analogues we have studied the ring-opening of the epoxide (**15**) with a variety of nucleophiles. Epoxide (**15**) has been reacted with methanolic sodium methoxide (Scheme 5), lithium aluminium hydride (Scheme 6), tris(dimethylamino)sulphur trimethylsilyldifluoride (TASF) (Scheme 7), and lithium dimethyl(cyano)copper (I) (Scheme 8). These results indicate that selectivity for nucleophilic attack at C6 over C1 is largely controlled by temperature. Selectivity was greatest with the cuprate (21:1) where the reaction was carried out at -30°C, and least for the dioxane alcohol (1.6:1) where the reaction was carried out at 100-110°C. Reactions carried out at intermediate temperatures gave intermediate selectivities. Nucleophilic attack at C6 has led to the synthesis of four novel trisphosphate analogues *via* a sequence of reactions involving debenzylation, phosphorylation and deprotection. These ring-opened products are also precursors to a variety of cyclitols, obtained by subsequent hydrogenolysis and treatment with moist trifluoroacetic acid (*13*) (Schemes 5-8). These derivatives could be used to probe the various enzyme mechanisms in the PI cycle, including binding and hydrolysis of phosphate groups.

To this point, we have shown the versatility of this novel approach to cyclitols and *myo*-inositol phosphates by a variety of ring-openings of the epoxide (**15**). A large number of structures and substitution patterns are therefore available,

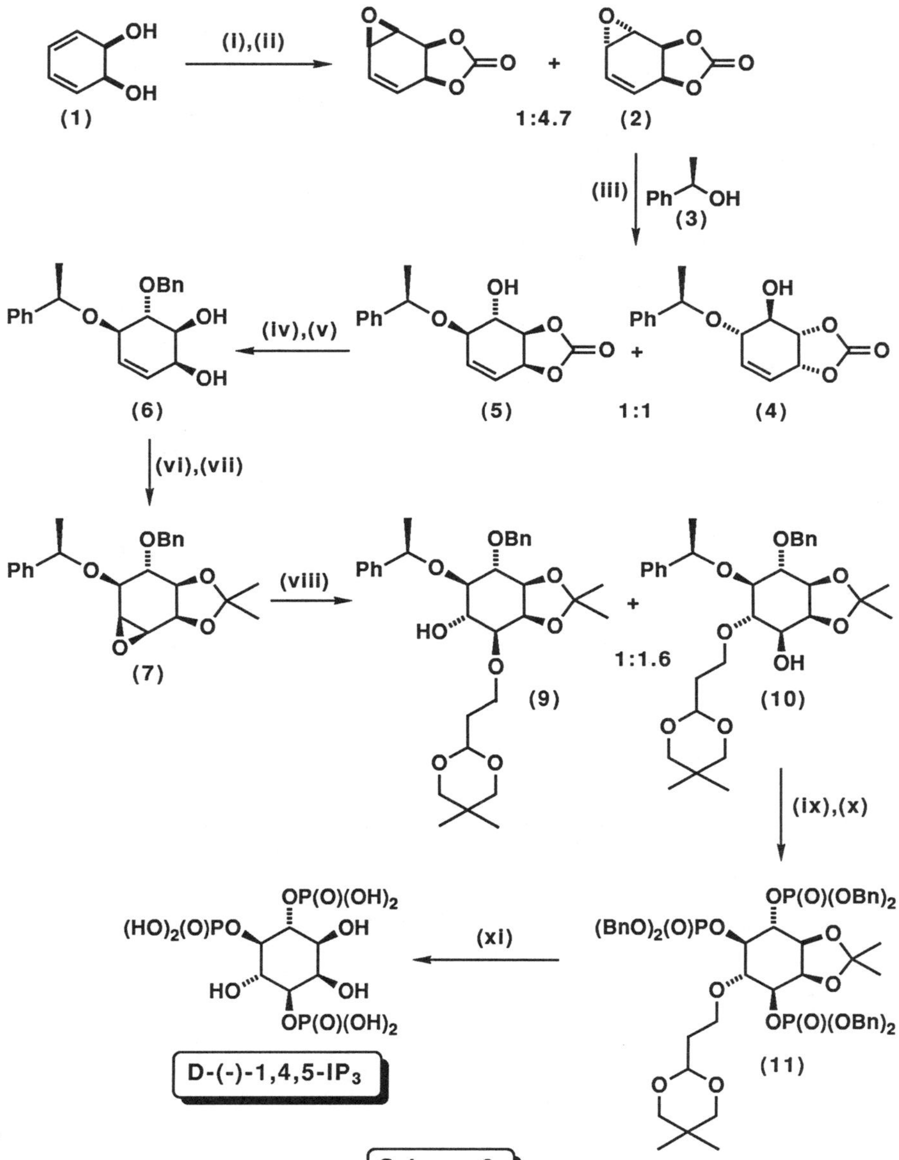

Scheme 3

(i) (MeO)$_2$CO, MeO$^-$Na$^+$, MeOH; (ii) *m*CPBA, DCM, 57% over 2 steps; (iii) cat. HBF$_4$.OEt$_2$, DCM, 67%; (iv) BnBr, Ag$_2$O, DMF, 3 days, 100%; (v) Et$_3$N/MeOH/H$_2$O, 3 days, 99%; (vi) *m*CPBA, DCM, 87%; (vii) 2,2-Dimethoxypropane, CSA, DCM, 89%; (viii) (8), NaH, TMEDA, 110°C, 3 days, 95%; (ix) H$_2$, 10% Pd-C, EtOH, 16 h, 100%; (x) ^{n}BuLi, ^{i}Pr$_2$NH, THF, tetrabenzylpyrophosphate, -30°C to RT, 67%; (xi) a. H$_2$, 10% Pd-C, EtOH, 48 h; b. 80% aq. TFA, 4h, 88% overall.

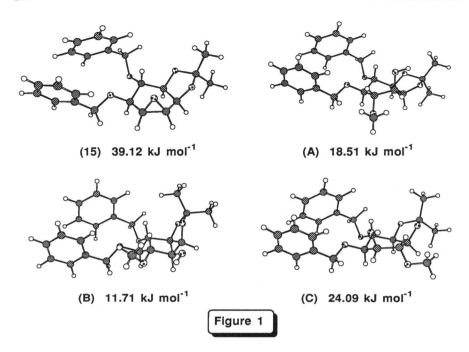

(15) 39.12 kJ mol-1
(A) 18.51 kJ mol-1
(B) 11.71 kJ mol-1
(C) 24.09 kJ mol-1
Figure 1

Figure 1. MM2 minimized structures of the nucleophilic ring-opening of epoxide (15) with methoxide anion.

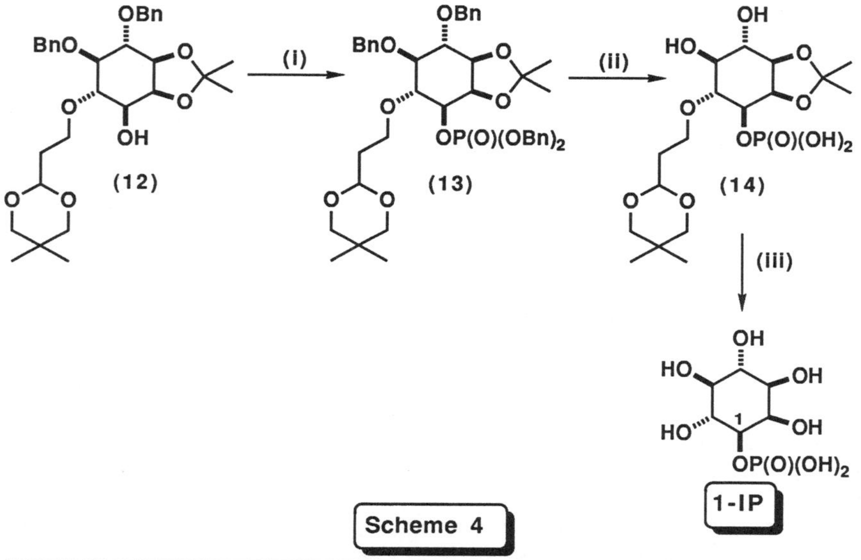

OBn
BnO
O
O
OH
(12)
(i)
OBn
BnO
OP(O)(OBn)2
(13)
(ii)
OH
HO
OP(O)(OH)2
(14)
(iii)
OH
HO
OH
HO
1
OH
OP(O)(OH)2
1-IP
Scheme 4

(i) ${}^{n}BuLi$, ${}^{i}Pr_2NH$, THF, tetrabenzylpyrophosphate, -30°C to RT, 57%; (ii) H_2, 10% Pd-C, EtOH, 100%; (iii) 80% aq. TFA, 4h, 21%.

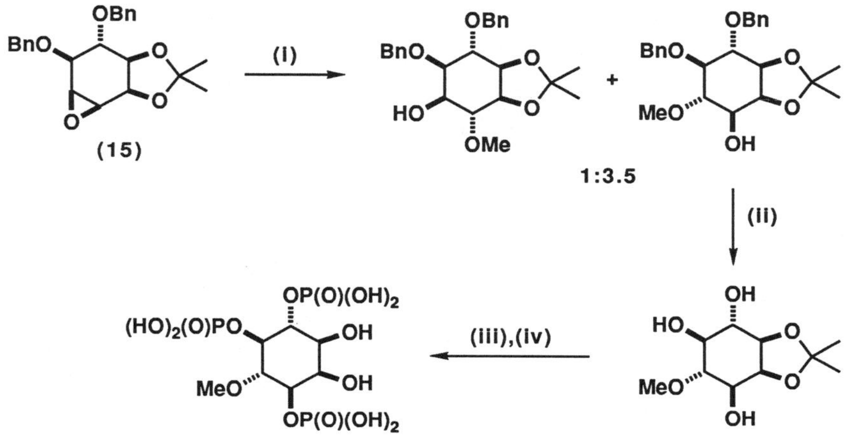

Scheme 5

(i) MeO⁻Na⁺, MeOH, reflux, 72h, 95%; (ii) H₂, 10% Pd-C, EtOH, 23h, 100%; (iii) ⁿBuLi, ⁱPr₂NH, THF, DMSO, tetrabenzylpyrophosphate, -30°C to RT, 61% ; (iv) a.TMSBr, DCM, 55 min; b. H₂O, 80 min, 62% overall.

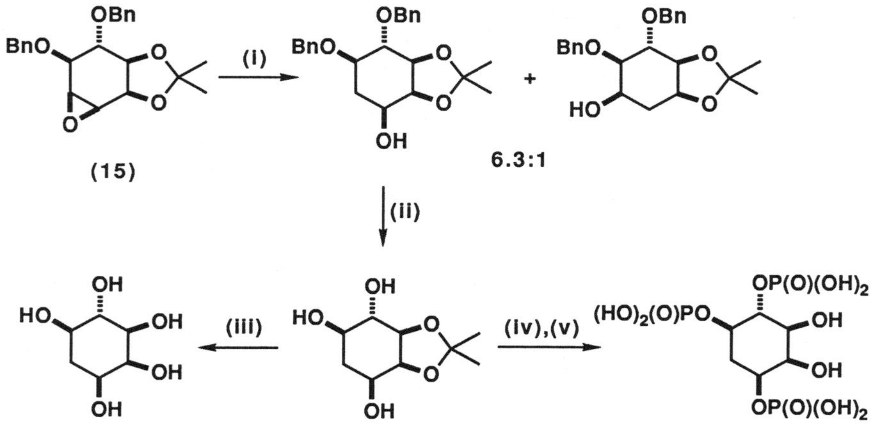

Scheme 6

(i) LiAlH₄, Et₂O, reflux, 2h, 88%;   (ii) H₂, 10% Pd-C, EtOH, 39h, 100%;   (iii) 80% aq. TFA, 4h, 97%; (iv) ⁿBuLi, ⁱPr₂NH, THF, tetrabenzylpyrophosphate, -30°C to RT, 56% ; (v) a. TMSBr, DCM, 65 min; b. H₂O, 80 min, 79% overall.

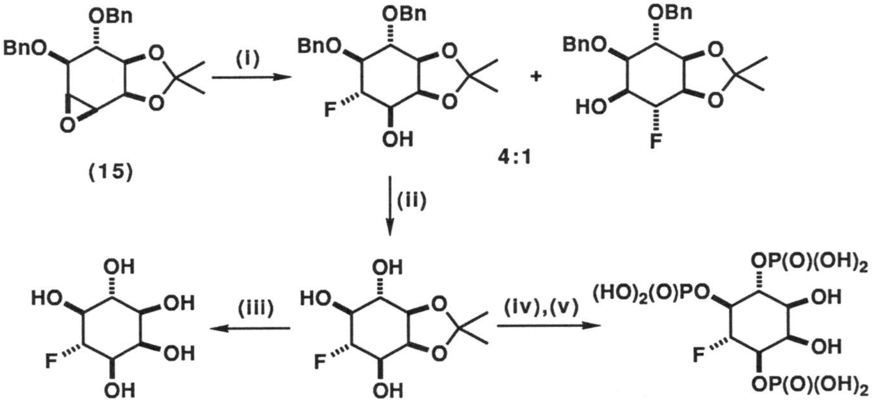

Scheme 7

(i) TASF, THF, reflux, 4 days, 74%;   (ii) H$_2$, 10% Pd-C, EtOH, 17h, 100%;   (iii) 80% aq. TFA, 4h, 100%;   (iv) ^{n}BuLi, ^{i}Pr$_2$NH, THF, tetrabenzylpyrophosphate, -30°C to RT, 77% ;   (v) a. TMSBr, DCM, 65 min; b. H$_2$O, 80 min, 93% overall.

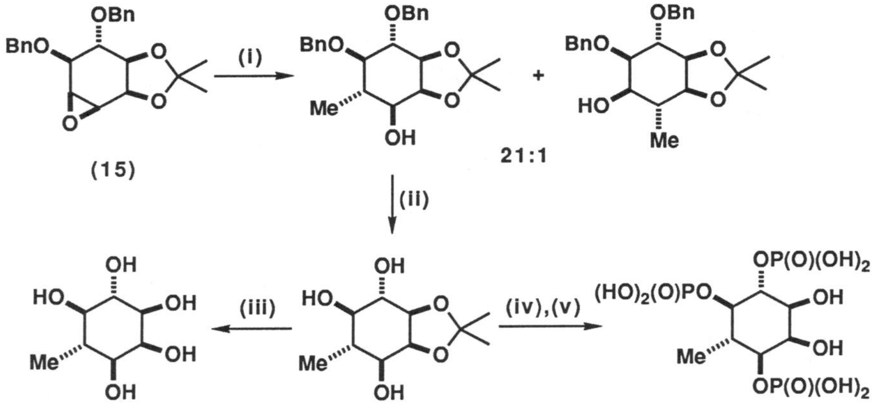

Scheme 8

(i) Me$_2$Cu(CN)Li$_2$, THF, -30°C, 22h, 74%;   (ii) H$_2$, 10% Pd-C, EtOH, 16h, 97%;   (iii) 80% aq. TFA, 5h, 81%;   (iv) ^{n}BuLi, ^{i}Pr$_2$NH, THF, tetrabenzylpyrophosphate, -30°C to RT, 41% ;   (v) a. TMSBr, DCM, RT, 65 min; b. H$_2$O, 80 min, 82% overall.

many of which would be difficult to obtain by standard modifications of *myo*-inositol itself.

A key building block (5) in the synthesis of D-(-)-IP₃ also paves the way to another natural product, (+)-conduritol F (*14*). This cyclohexenetetrol, found in most green plants, is obtained by simple deprotection of alcohol (5) with sodium in liquid ammonia (eq.2). Thus, the four contiguous asymmetric carbon atoms in (+)-conduritol F are established in only five steps from benzene, again utilizing *Pseudomonas putida* oxidation to introduce the necessary *cis*-1,2-diol functionality. The unnatural antipode, (-)-conduritol F, was prepared in a similar fashion in 85% yield from (4). Derivatives of these conduritols are of particular interest as potential inhibitors for glycosidases (*28-29*).

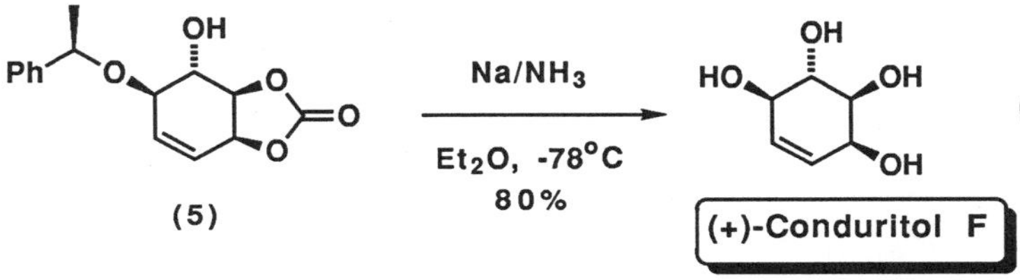

(eq.2)

Intermediate (10), with its differential hydroxyl protection, facilitates ready access to a series of compounds based on the phospholipid phosphatidylinositol 4,5-bisphosphate (PIP₂). However, the major synthetic challenge in this area is PIP₂ itself with its highly unsaturated acyl side chain (Scheme 9).

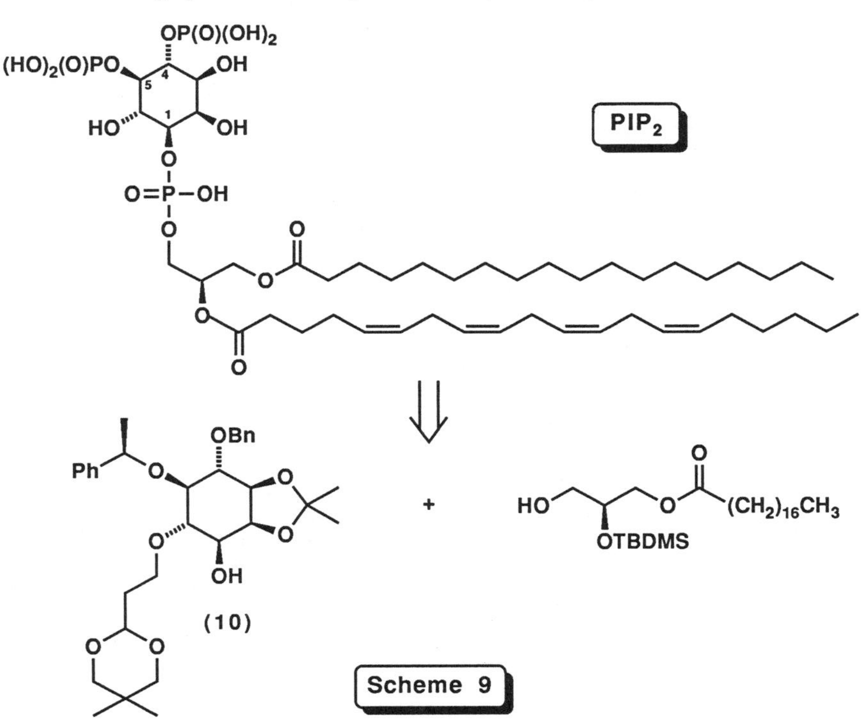

Scheme 9

Our approach involves protection of (10) as its silyl ether (16), hydrogenolysis, and then phosphorylation with tetrabenzylpyrophosphate to yield (17) in 83%. Deprotection of the bisphosphate (17) with tetra-n-butylammonium fluoride furnishes (18) which is a suitable inositol building block for the PIP$_2$ synthesis (Scheme 10).

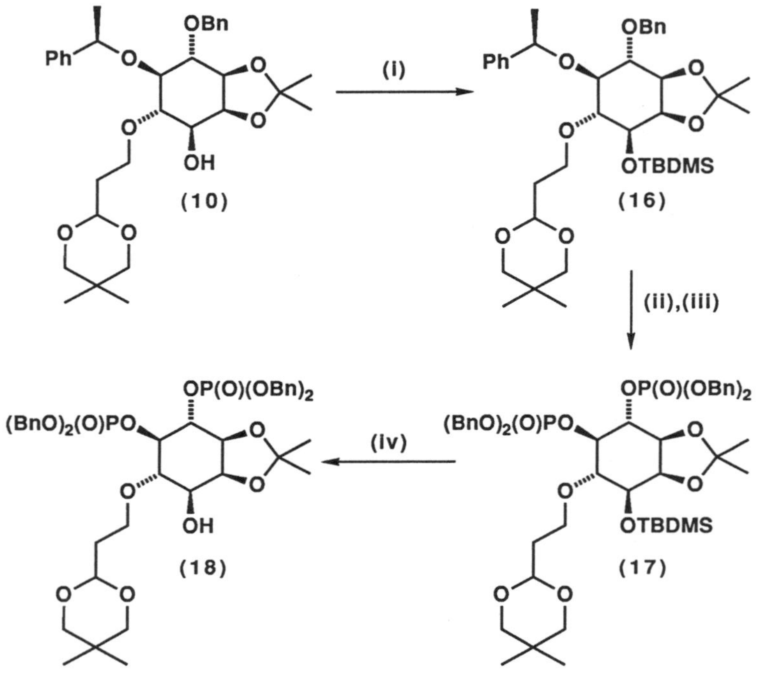

Scheme 10

(i) TBDMSCl, Imidazole, DMF, 65°C, 97%; (ii) H$_2$, Pd-C; (iii) ^{n}BuLi, ^{i}Pr$_2$NH, TBPP, THF, -30°C, 83% over 2 steps; (iv) TBAF, THF, 99%.

Since naturally occurring PIP$_2$ contains a mixed 1,2-diacyl glycerol fragment with a highly unsaturated sn-2-acyl chain, entry to this compound via known synthetic methods has been severely restricted (30). The major problem is choice of protecting groups for there is a propensity for deleterious 1,2-acyl migration to occur, as well as the highly unsaturated acyl chain suffering facile oxidation. In our approach to the 1,2-substituted-sn-glyceride we started with 1,2-O-isopropylideneglycerol (19), and by a series of standard procedures obtained the mono-substituted saturated glyceride (20) without any functional group migration (Scheme 11).

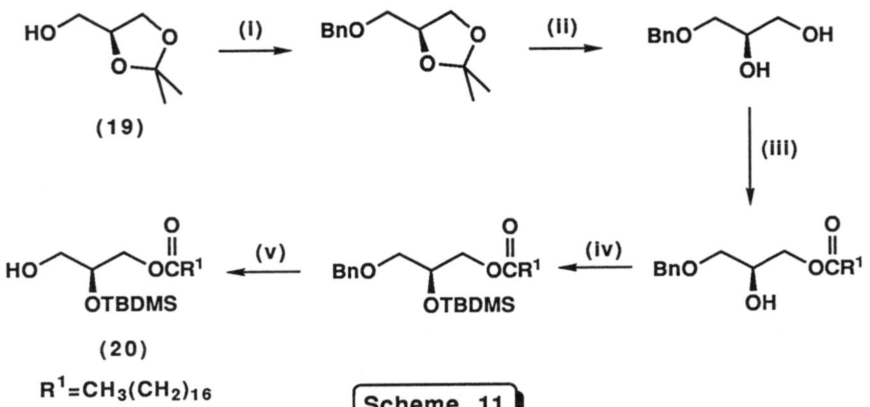

Scheme 11

(i) a. NaH; b. BnBr, 97%; (ii) CSA, MeOH, 59%; (iii) R^1CO_2H, DCC, DMAP, 0°C, 55%;
(iv) TBDMSCl, Imid., 60°C, 97%; (v) H_2, Pd-C, 100%.

It is then proposed to form the phosphodiester linkage between the glyceride
(20) and the *myo*-inositol derivative (18) in PIP_2 *via* a phosphite coupling approach
(*30-33*). Deprotection of the 2-position of the glyceride under normal conditions will
hopefully avoid migration of the functional groups since the thermodynamically
stable 1,3-substituted product will be formed. Introduction of the arachidonic acid
group at the 2-position of the glyceride followed by removal of all the protecting
groups will access PIP_2.

Owing to current interest in chiral cyclohexadiene diols as novel starting
materials for organic synthesis, we also report here several reactions of *cis*-3-fluoro-
3,5-cyclohexadiene-1,2-diol (21) which may well find applications in the future. The
cis-diol functionality can be protected with a variety of protecting groups, including
acetonide and carbonate. The diene can then be stereo- and regio-selectively
epoxidised using *m*-chloroperoxybenzoic acid or dimethyldioxirane (Scheme 12).

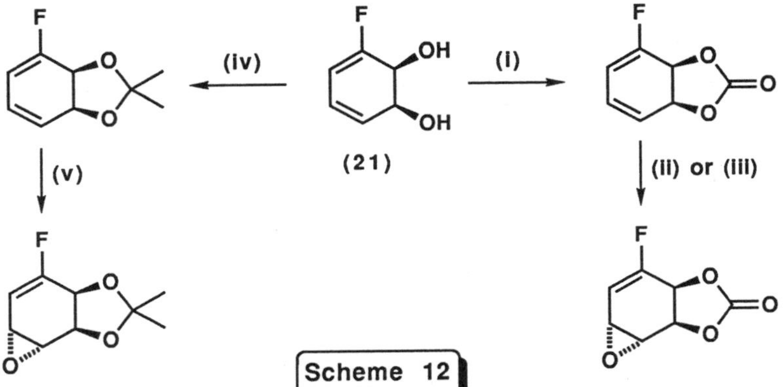

Scheme 12

(i) CDI, DCM, 0°C, 82%; (ii) *m*CPBA, DCM, 0°C to RT, 36%; (iii) Dimethyldioxirane,
0°C, 20%; (iv) 2,2-Dimethoxypropane, CSA, DCM, 75%; (v) *m*CPBA, pH 8, DCM, 51%.

This novel approach for the conversion of arenes to polyol derivatives, as illustrated by the synthesis of D-(-)-IP$_3$, harnessing biotechnology and organic synthesis is a powerful strategy for future development. This is especially true when the biotechnological reaction cannot be readily achieved by conventional synthetic methods. Indeed, it would be profitable to seek out these unique combinations and apply them to other situations.

Acknowledgments. We thank the SERC, ICI Strategic Research Fund and The Croucher Foundation (L.L.Y.) for financial support. We also thank Dr. S.C. Taylor (ICI Biological Products) for generous supplies of (**1**) and (**21**).

References.
1. Brockerhoff, H.; Ballou, C.E. *J. Biol. Chem.* **1961**, *236*, 1907.
2. Michell, R.H. *Biochim. Biophys. Acta* **1975**, *415*, 81.
3. Berridge, M.J.; Irvine, R.F. *Nature* **1984**, *312*, 315.
4. Berridge, M.J.; Irvine, R.F. *Nature* **1989**, *341*, 197.
5. Billington, D.C. *Chem. Soc. Rev.* **1989**, *18*, 83.
6. Potter, B.V.L. *Nat. Prod. Rep.* **1990**, 7, 1.
7. Berridge, M.J.; Michell, R.H. *Philos. Trans. R. Soc. Lond. B.* **1988**, *320*, 235.
8. Gibson, D.T.; Koch, J.R.; Kallio, R.E. *Biochemistry* **1968**, 7, 2653.
9. Ley, S.V.; Sternfeld, F.; Taylor, S. *Tetrahedron Lett.* **1987**, *28*, 225.
10. Ley, S.V.; Sternfeld, F. *Tetrahedron* **1989**, *45*, 3463.
11. Ley, S.V.; Sternfeld, F. *Tetrahedron Lett.* **1988**, *29*, 5305.
12. Ley, S.V.; Parra, M.; Redgrave, A.J.; Sternfeld, F.; Vidal, A. *Tetrahedron Lett.* **1989**, *30*, 3557.
13. Ley, S.V.; Parra, M.; Redgrave, A.J.; Sternfeld, F. *Tetrahedron* **1990**, *46*, 4995.
14. Ley, S.V.; Redgrave, A.J. *Synlett* **1990**, 393.
15. Ley, S.V. *Pure & Appl. Chem.* **1990**, *62*, 2031.
16. Nakajima, M.; Tomida, I.; Takei, S. *Chem. Ber.* **1959**, *92*, 163.
17. Hudlicky, T.; Seoane, G.; Pettus, T. *J. Org. Chem.* **1989**, *54*, 4239.
18. Hudlicky, T.; Price, J.D.; Luna, H.; Andersen, C.M. *Synlett* **1990**, 309.
19. Howard, P.W.; Stephenson, G.R.; Taylor, S.C. *J. Chem. Soc. Chem. Commun.* **1988**, 1603.
20. Carless, H.A.J.; Billinge, J.R.; Oak, O.Z. *Tetrahedron Lett.* **1989**, *30*, 3113.
21. Namata, A.; Hokimoto, K.; Shimada, A.; Yamaguchi, H.; Takaishi, K. *Chem. Pharm. Bull.* **1979**, *27*, 602.
22. Reece, J.C.; Chan, B.G.; Waiss Jr., A.C. *J. Chem. Ecol.* **1982**, *8*, 1429.
23. Drewer, D.L.; Binder, R.G.; Chan, B.G.; Waiss Jr., A.C.; Hartwig, E.E.; Beland, G.L. *Experientia* **1979**, *35*, 1182.
24. Narayanan, C.R.; Joshi, D.D.; Miyumdar, A.M.; Dhekne, V.V. *Curr. Sci.*, **1987**, *56*, 139.
25. Using the MM2 forcefield in the program Macromodel v 3.0, W.C. Still, Columbia University, New York, USA.
26. Khorana, H.G.; Todd, A.R. *J. Chem. Soc.* **1953**, 2257.
27. Watanabe, Y.; Nakahira, H.; Bunya, M.; Ozaki, S. *Tetrahedron Lett.* **1987**, *28*, 4179.
28. Legler, G.; Bause, E. *Carbohydr. Res.* **1973**, *28*, 45.
29. Legler, G.; Loth, W. *Hoppe-Seyler's Z. Physiol. Chem.* **1973**, *354*, 243.
30. Matin, S.F.; Josey, J.A. *Tetrahedron Lett.* **1988**, *29*, 3631.
31. Bruzik; K.S.; Salamonczyk, G.; Stec, W.J. *J. Org. Chem.* **1986**, *51*, 2368.
32. Dreef, C.E.; Elie, C.J.J.; Hoogerhout, P.; van der Marel, G.A.; van Boom, J.H. *Tetrahedron Lett.* **1988**, *29*, 6513.
33. Inami, K.; Teshima, T.; Emura, J.; Shiba, T. *Tetrahedron Lett.* **1990**, *31*, 4033.

RECEIVED February 11, 1991

Chapter 11

Enantiospecific Synthesis of Inositol Polyphosphates, Phosphatidylinositides, and Analogues from (−)-Quinic Acid

J. R. Falck and Abdelkrim Abdali

Department of Molecular Genetics and Pharmacology, University of Texas Southwestern Medical Center, Dallas, TX 75235

An efficient and versatile strategy for the preparation of highly differentiated cyclitols from (-)-quinic acid was developed and exploited for the enantiospecific total synthesis of representative inositol polyphosphates and phosphatidylinositides. The 5-methylenephosphonate analogue of D-*myo*-inositol 1,4,5-trisphosphate was also prepared and shown to be a long-lived agonist of calcium mobilization.

In recognition of the physiologic significance of phosphorylated cyclitols and their limited availability from natural sources, there has been a resurgence of interest in the chemical synthesis of functionalized inositols and analogues (*1,2*). Until recently, virtually all preparative studies utilized *myo*-inositol as the initial precursor and relied upon sometimes circuitous protection/deprotection sequences to differentiate amongst this cyclitol's six hydroxyls. Furthermore, since the *myo*-isomer is a meso form of inositol, a resolution is required to obtain optically active products. These limitations have been partially addressed (*3-5*) using pinitol and quebrachitol, two chiral, but not widely available natural cyclitols. Efforts to exploit carbohydrates have also been reported (*6*). It should be noted that the novel intermediate, *cis*-1,2-dihydroxycyclohexa-3,5-diene, arising from microbial oxidation of benzene, has also been used as the starting point for inositol syntheses (*7,8*). Given the increasing popularity of enzymatic methodology in organic synthesis, it is reasonable to anticipate additional contributions utilizing such procedures.

As part of a comprehensive program in these laboratories for the stereorational total synthesis of inositol polyphosphates and associated phospholipids, we developed a conceptually different, and potentially more flexible, approach to functionalized cyclitols that exploits a relatively inexpensive and readily available member of the chiral pool, (-)-quinic acid (**1**). Our strategy also offers, in principle, greater latitude for preparing configurational and constitutional analogues.

0097–6156/91/0463–0145$06.00/0

Inositol 1,4,5-Trisphosphate

Our initial synthetic target was D-*myo*-inositol 1,4,5-trisphosphate (**10**), the calcium-mobilizing intracellular second messenger of the phosphatidylinositol (PI) cycle (*9*). Of the possible perspectives from which to view **1** with respect to the *myo*-inositol moiety **2** present in **10**, the orientation depicted in **1a** provides the greatest topologic congruity, i.e., the three contiguous quinate alcohols have the same relative and absolute configurations as C(2), C(1), and C(6), respectively, when **1a** is superimposed upon **2**. Projection of orientation **1b** reveals coincidence with only two stereocenters, those at C(2) and C(3). The third alcohol at C(1) is epimeric. Yet, despite its lesser stereochemical homology, other considerations (*vide infra*) made **1b** the more attractive candidate provided satisfactory solutions could be found for several outstanding issues. These included: (i) differentiation of the hydroxyls, (ii) stereospecific oxidation of both methylenes corresponding to C(4) and C(6), and (iii) one-carbon degradation to remove the carboxyl. The realization (*10*) of these objectives culminating in an efficient total synthesis of **10** is outlined in Scheme I.

Firstly, the C(1)-stereochemistry was adjusted using a modification of literature procedure that improved the overall yield and simplified isolation of intermediates. Specifically, (-)-quinic acid (**1**) was converted to lactone **3** by concurrent lactonization/ ketalization utilizing cyclohexanone and Amberlite IR-120 resin followed by mesylation of the tertiary alcohol under standard conditions. Sequential lactone methanolysis, pyridinium chlorochromate (PCC) oxidation, and Et_3N induced mesylate elimination generated an enone from which ester **4** was obtained by hydride delivery exclusively from the less hindered ß-face.

Taking advantage of the residual functionality created above, the C(6)-hydroxyl was introduced stereospecifically with the desired configuration. For this, ester **4** was transformed to phenyl selenide **5** by protection of the C(1)-alcohol as its ß-trimethylsilylethoxymethyl(SEM) ether, reduction of the ester using diisobutylaluminum hydride (DIBAL-H) at -78°C, and selenylation of the resultant primary alcohol with N-(phenylseleno)phthalimide in the presence of Bu_3P (*11*). Stereoselective *in situ* [2,3]-sigmatropic rearrangement of the allylic selenoxide derived from **5** and benzylation of the product evolved **6** as the sole product. Due to the high cost of the selenating reagent, the sulfide version of **5** was also examined. In this instance, peracid oxidation yielded an *ca.* 1:1 mixture of sulfoxides that led in refluxing benzene to a mixture of epimeric alcohols (α/ß 1:3); at 45°C, however, only the ß-isomer was observed. The favorable outcome of this rearrangement can be attributed to (i) facile racemization of the sulfoxide via rapid equilibration with its sulfinate ester and (ii) preferential approach of the thiophile from the ß-face. Inspection of molecular models confirms the α-side is well-screened by the bulky cyclohexylidene substituent (cf.**20**).

To functionalize the remaining methylene, **6** was subjected to SeO_2 allylic oxidation (78%), then ozonolysis (83%) in $MeOH/CH_2Cl_2$. However, hydride

Scheme I

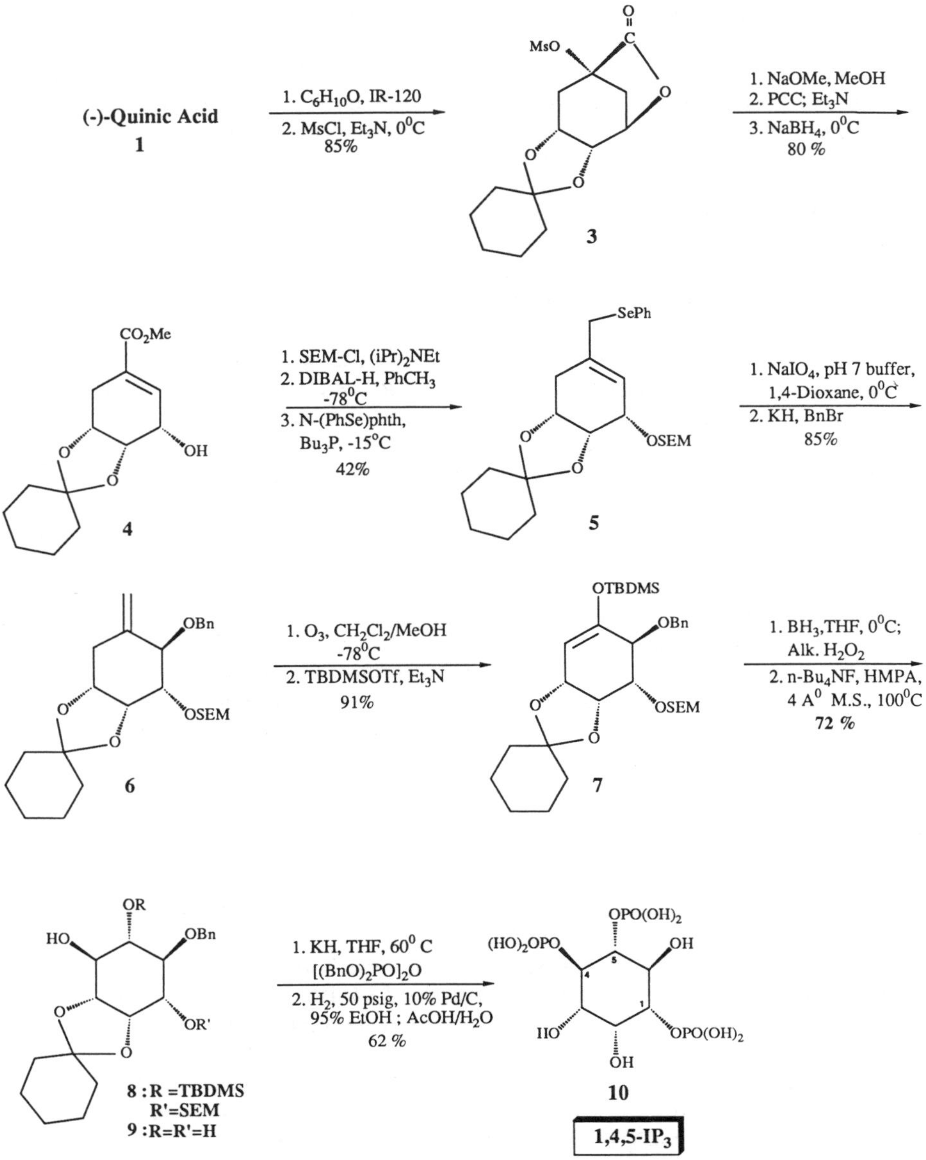
(-)-Quinic Acid
1
1. C6H10O, IR-120
2. MsCl, Et3N, 0°C
85%
MsO
3
1. NaOMe, MeOH
2. PCC; Et3N
3. NaBH4, 0°C
80 %
CO2Me
OH
4
1. SEM-Cl, (iPr)2NEt
2. DIBAL-H, PhCH3
-78°C
3. N-(PhSe)phth,
Bu3P, -15°C
42%
SePh
OSEM
5
1. NaIO4, pH 7 buffer,
1,4-Dioxane, 0°C
2. KH, BnBr
85%
OBn
OSEM
6
1. O3, CH2Cl2/MeOH
-78°C
2. TBDMSOTf, Et3N
91%
OTBDMS
OBn
OSEM
7
1. BH3,THF, 0°C;
Alk. H2O2
2. n-Bu4NF, HMPA,
4 A° M.S., 100°C
72 %
OR
HO
OBn
OR'
8 : R =TBDMS
R'=SEM
9 : R=R'=H
1. KH, THF, 60° C
[(BnO)2PO]2O
2. H2, 50 psig, 10% Pd/C,
95% EtOH ; AcOH/H2O
62 %
OPO(OH)2
(HO)2OPO
OH
HO
OPO(OH)2
OH
10
1,4,5-IP3

reduction of the ketone 11, thus obtained, afforded a separable mixture of C(5)-diastereomeric alcohols under a variety of conditions. Efforts to epimerize the undesired isomer proved disappointing. In contrast, ozonolysis of 6 and conversion to silyl enol ether 7 using excess *tert*-butyldimethylsilyl (TBDMS) triflate according to Corey (12) proceeded smoothly and with virtually complete regiospecificity. Hydroboration of the enol ether from the more accessible ß-face followed by careful alkaline peroxide oxidation furnished differentially protected cyclitol 8. While the TBDMS ether was easily severed, more drastic conditions (n-Bu$_4$NF,HMPA,100°C) were needed to coerce the SEM in order to obtain triol 9. The identity of 9 was confirmed by comparisons of its physical [mmp 136-137°C; lit. (13) 137-139°C] and spectral characteristics with an authentic sample.

Final elaboration of 9 with tetrabenzyl pyrophosphate and removal of the protecting groups as described by Vacca (14) provided D-*myo*-inositol 1,4,5-trisphosphate 10, isolated as its hexasodium salt.

5-Methylenephosphonate Analogue of 1,4,5-IP$_3$

In vivo, 1,4,5-IP$_3$ (10) is rapidly metabolized by either of two divergent pathways: (a) initial phosphorylation by a 3-kinase and subsequent cleavage of the C(5)-phosphate, or (b) direct dephosphorylation by a specific 5-phosphatase (1,2). Intervention in these pathways may provide new insights into the function and possible therapeutic manipulation of the PI cycle. Special emphasis has been placed on the development of modified inositol polyphosphates with metabolically more stable phosphorus functionalities (15-19). Consequently, it was of interest to prepare 14, the 5-methylenephosphonate analogue (20) of 1,4,5-IP$_3$, by an extension of our strategy that utilizes both the chirality and *complete* carbon framework of (-)-quinic acid (Scheme II).

Fluoride ion induced desilylation of 6 in hexamethylphosphoric triamide (HMPA) and kinetically (21) controlled addition of phenylselenenyl bromide across the exocyclic alkene provided mainly the anti-Markovnikov adduct. The latter underwent regiospecific oxidative elimination to the somewhat labile allylic bromide 12 that was sufficiently pure for Michaelis-Becker phosphorylation (22) using sodium dibenzyl phosphite (generated *in situ* from NaH and dibenzyl phosphite) in the presence of 18-crown-6 to improve the solubility of the reagent in toluene. Subsequent hydroboration of the alkene followed by oxidative work-up using peracid, rather than alkaline H$_2$O$_2$ which hydrolyzes the phosphonate, furnished diol 13 with the desired all-*trans* configuration between the substituents at C(3)-C(6) (^{1}H NMR analysis: COSY, J-resolved). Introduction of the C(1) and C(4) phosphates by the two-step phosphite method (3) of Tegge and Ballou and removal of the protecting groups afforded 14, isolated as its sodium salt.

Phosphonate 14 elicits contraction of bovine tracheal smooth muscle (23), permeabilized with saponin, with an efficacy five to ten times less than a comparable concentration of 1,4,5-IP$_3$. Evaluation (24) of its ability to displace [^{32}P]-1,4,5-IP$_3$ bound to bovine adrenocortical membranes showed 14 has an affinity for the 1,4,5-IP$_3$ binding site *ca.* two orders of magnitude less than unlabeled 1,4,5-IP$_3$ (Figure 1). For release of Ca^{2+} from bovine adrenal

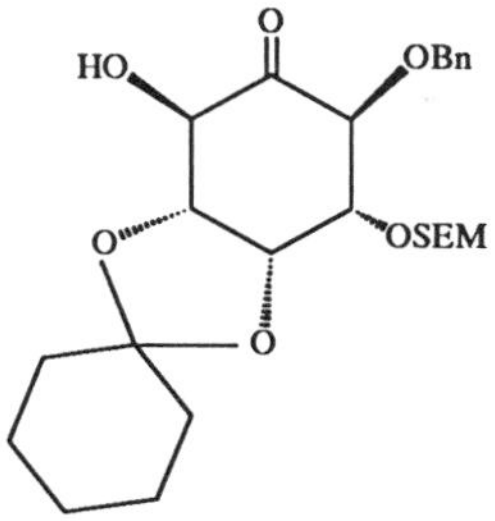

HO
O
OBn
OSEM
O
O

11

Scheme II

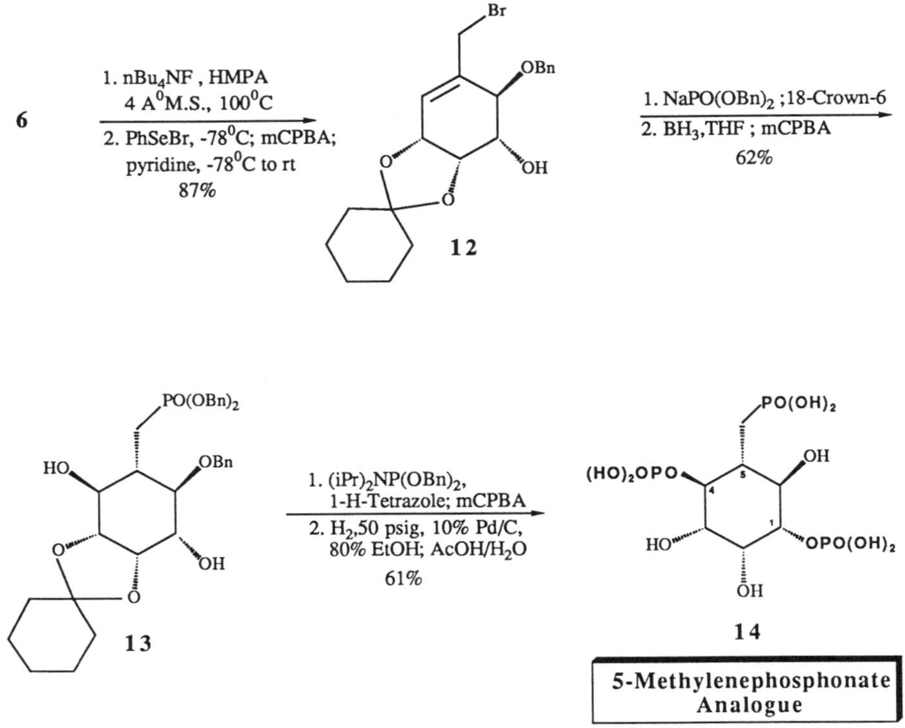

6
1. nBu4NF , HMPA
4 A0M.S., 100°C
2. PhSeBr, -78°C; mCPBA;
pyridine, -78°C to rt
87%
Br
OBn
OH
12
1. NaPO(OBn)2 ;18-Crown-6
2. BH3,THF ; mCPBA
62%
PO(OBn)2
HO
OBn
OH
13
1. (iPr)2NP(OBn)2,
1-H-Tetrazole; mCPBA
2. H2,50 psig, 10% Pd/C,
80% EtOH; AcOH/H2O
61%
PO(OH)2
(HO)2OPO
OH
HO
OPO(OH)2
OH
14
5-Methylenephosphonate
Analogue

microsomal preparations (24), as measured by Fura 2 fluorescence, 14 and 2,4,5-IP$_3$ were nearly equally potent and about one-fifth as active as the natural second messenger, 1,4,5-IP$_3$ (Figure 2). In contrast with the latter compound, 14 and 2,4,5-IP$_3$ give an initial sharp increase in the unbound Ca^{2+} concentration that then remains elevated above basal values (24). These data are consistent with an attenuated, but long-lived, 1,4,5-IP$_3$ agonist.

A Second Generation Approach

To date, almost two dozen inositol phosphate and phosphatidylinositide isomers have been isolated and identified from natural sources (1,2). Although recent investigations have established a complex interrelationship amongst these metabolites, the physiological role(s) for all but a few remains obscure. The more recent discovery (25,26) of a metabolically distinct class of 3-phosphorylated phosphatidylinositides portends even greater diversity. To help encompass this growing synthetic challenge, we sought to extend the scope of our basic theme by preparing a more highly differentiated intermediate and then illustrating its usefulness in the synthesis of some representative examples of current interest.

A convenient point of departure was lactone 15 available (27) on a large scale in 50-60% yield from 1 by vacuum sublimation (equation 1). The corresponding cyclic dibutyl stannylene ether, generated (28) in situ from 15, was enticed by an equivalent amount of n-Bu$_4$NI to react with SEM-Cl at 60°C to give diol 16 (78%). None of the regioisomeric SEM ether could be detected by chromatographic analysis. Yet, introduction of a benzyl or silyl protecting group at the adjacent alcohol proved unexpectedly troublesome. Unacceptable product mixtures were obtained due to the high reactivity of the tertiary alcohol. This approach was ultimately abandoned for one that called into service enone 17, an intermediate in the preceding syntheses.

Concurrent reduction of the ester and keto functionalities in 17 with DIBAL-H followed by selective replacement of the primary alcohol using phenyl disulfide/tributylphosphine was rewarded by a good yield of phenylsulfide 18 (Scheme III). A 3,4-dimethoxybenzyl was staked onto the remaining secondary alcohol and the cyclohexylidene was removed under mild acidic conditions. Tin mediated etherification of the resultant diol 19 with SEM-Cl using nBu$_4$NI (vide supra) showed lower regioselectivity (4:1) for the equatorial C(3)-alcohol than did CsF catalysis (24:1). Subsequent benzylation leading to 20 was uneventful. Low temperature peracid oxidation gave a ~ 1:1 mixture of sulfoxides that was transformed into olefin 21, free of any α-isomer, by [2,3]-sigmatropic rearrangement and benzylation of the newly generated allylic alcohol. In this instance, rearrangement required a significantly longer reaction time and bulkier thiophile than the comparable cyclohexylidene examples described above. Having observed that the 3,4-dimethoxybenzyl ether did not survive ozonolytic cleavage of the exocyclic olefin, ketone 22 was secured by an alternative two-step, but very efficient, sequence involving OsO$_4$ glycolization and Pb(OAc)$_4$ cleavage. Proceeding with 22 along the now well established route of enol ether

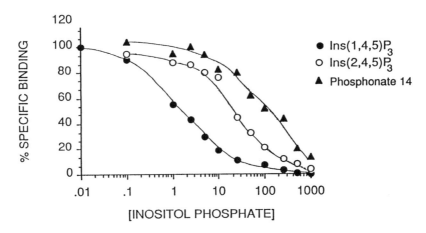

Figure 1. ^{32}P-Ins(1,4,5)P$_3$ displacement in bovine adrenal microsomal membranes.

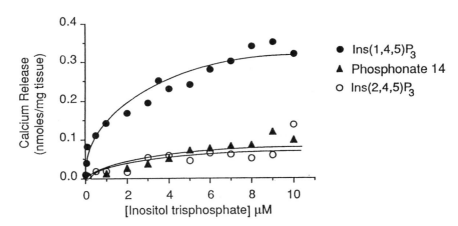

Figure 2. Fura 2 measurement of calcium release in bovine adrenal microsomes.

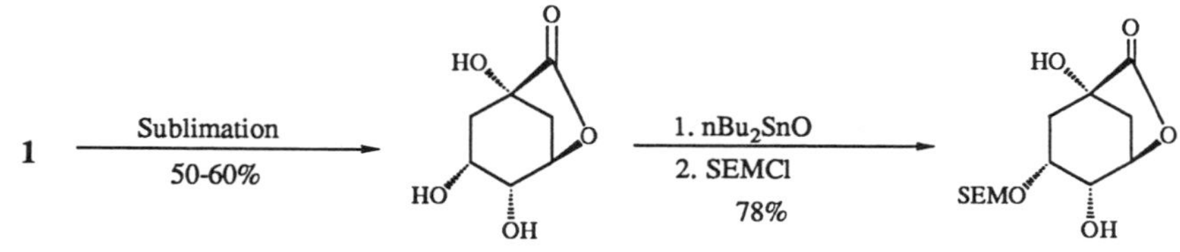

$$1 \xrightarrow[\text{50-60\%}]{\text{Sublimation}} \mathbf{15} \xrightarrow[\substack{\text{2. SEMCl} \\ 78\%}]{\text{1. nBu}_2\text{SnO}} \mathbf{16} \quad (1)$$

Scheme III

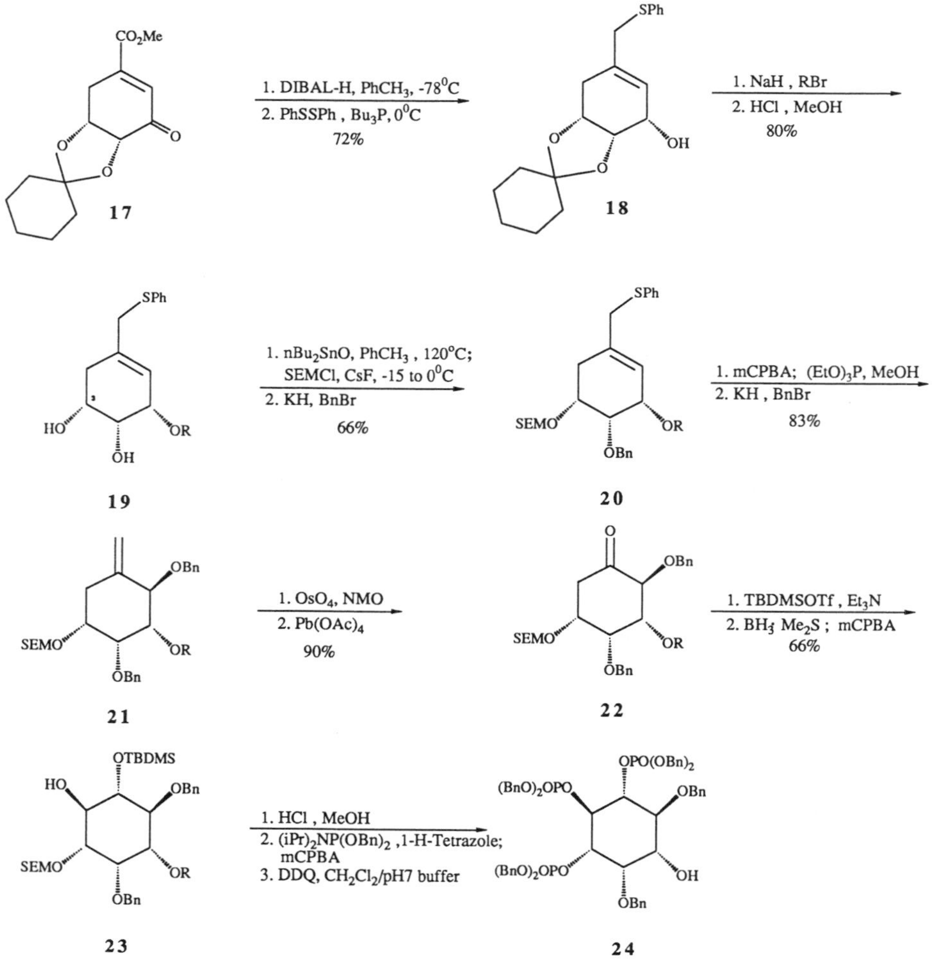

R= 3,4-(MeO)$_2$C$_6$H$_3$CH$_2$-

Scheme IV

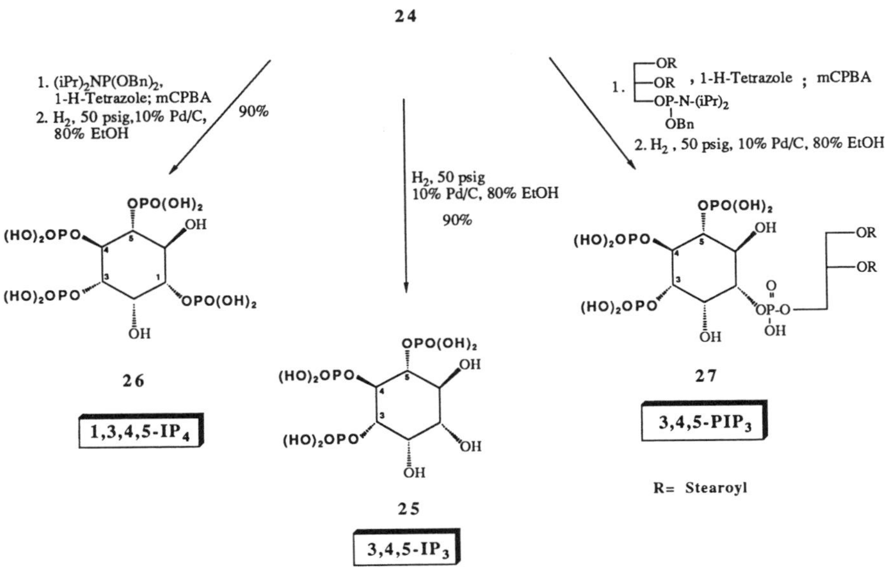

24

1. (iPr)$_2$NP(OBn)$_2$,
1-H-Tetrazole; mCPBA
2. H$_2$, 50 psig,10% Pd/C,
80% EtOH 90%

1. ┌─OR
│─OR , 1-H-Tetrazole ; mCPBA
└─OP-N-(iPr)$_2$
OBn
2. H$_2$, 50 psig, 10% Pd/C, 80% EtOH

H$_2$, 50 psig
10% Pd/C, 80% EtOH
90%

OPO(OH)$_2$
(HO)$_2$OPO ⟋ OH
(HO)$_2$OPO ⟍ OPO(OH)$_2$
OH

26

1,3,4,5-IP$_4$

OPO(OH)$_2$
(HO)$_2$OPO ⟋ OH
(HO)$_2$OPO ⟍ OH
OH

25

3,4,5-IP$_3$

OPO(OH)$_2$
(HO)$_2$OPO ⟋ OH ┌─OR
│─OR
(HO)$_2$OPO ⟍ O
‖
OP-O─┘
OH OH

27

3,4,5-PIP$_3$

R = Stearoyl

formation, hydroboration and oxidative work-up with peracid (basic hydrogen peroxide induced migration and/or loss of the t-BuMe$_2$Si protecting group) gave rise to the fully hydroxylated and highly differentiated cyclitol 23. Consecutive hydrolysis of the SEM and silyl ethers by methanolic hydrogen chloride, phosphorylation of the C(3), (4), and (5)-alcohols, and DDQ promoted cleavage of the 3,4-dimethoxybenzyl ether afforded 24, the pivotal intermediate for the series of polyphosphorylated inositols shown in Scheme IV.

Conventional catalytic debenzylation of 24 gave 3,4,5-IP$_3$ (25). This isomer has not been isolated yet from natural sources, but is of considerable interest for current structure-activity studies. Decoration of 24 with an additional phosphate prior to deprotection evolved 1,3,4,5-IP$_4$ (26), a metabolite of 10 that is thought to regulate extracellular calcium influx (1,2). Union of 24 with a 1,2-di-O-stearoyl-sn-glycero-3-phosphoramidite (29), mCPBA oxidation of the intermediate phosphite triester, and catalytic hydrogenolysis produced 3,4,5-PIP$_3$ (27), the product of a unique PI-3-kinase that may play an important role in cellular proliferation (25,26).

The development of additional synthetic cyclitols suitable for the preparation of glycosyl-phosphatidylinositides (30) including those with the rare *chiro*-inositol subunit are in progress.

Acknowledgment: Work from the authors' laboratories was supported by the American Heart Association and the Robert A. Welch Foundation.

Literature Cited

(1) Billington, D.C. *Chem. Soc. Rev.* **1989**, *18*, 83-122.
(2) Potter, B.V.L. *Nat. Prod. Reports* **1990**, *7*, 1-24.
(3) Tegge, W.; Ballou, C.E. *Proc. Natl. Acad. Sci. USA* **1989**, *86*, 94-98.
(4) Akiyama, T.; Takechi, N.; Ozaki, S. *Tetrahedron Lett.* **1990**, *31*, 1433-1434.
(5) Kozikowski, A.P.; Fauq, A.H.; Rusnak, J.M. *Tetrahedron Lett.* **1989**, *30*, 3365-3368.
(6) Watanabe, Y.; Mitani, M.; Ozaki, S. *Chem. Lett.* **1987**, 123-126.
(7) Ley, S.V.; Sternfeld, F. *Tetrahedron Lett.* **1988**, *29*, 5305-5308.
(8) Carless, H.A.J.; Billings, J.R.; Oak, O.Z. *Tetrahedron Lett.* **1989**, *30*, 3113-3116.
(9) Downes, C.P. *Biochem. Soc. Trans.* **1989**, *17*, 259-268.
(10) Falck, J.R.; Yadagiri, P. *J. Org. Chem.* **1989**, *54*, 5851-5852.
(11) Grieco, P.A.; Jaw, J.Y.; Claremon, D.A.; Nicolaou, K.C. *J. Org. Chem.* **1981**, *46*, 1215-1217.
(12) Corey, E.J.; Cho, H.; Rucker, C.; Hua, D.H. *Tetrahedron Lett.* **1981**, *22*, 3455-3458.
(13) Vacca, J.P.; deSolms, S.J.; Huff, J.R. *J. Am. Chem. Soc.* **1987**, *109*, 3478-3479.
(14) Vacca, J.P.; deSolms, S.J.; Huff, J.R.; Billington, D.C.; Baker, R.; Kulagowski, J.J.; Mawer, I.M. *Tetrahedron* **1989**, *45*, 5679-5702.
(15) Cooke, A.M.; Gigg, R.; Potter, B.V.L. *J. Chem. Soc., Chem. Commun.* **1987**, 1525.
(16) Dreef, C.E.; van der Marel, G.A.; van Boom, J.H. *Recl. Trav. Chim. Pays-Bas* **1987**, *106*, 512.
(17) Cooke, A.M.; Noble, N.J.; Gigg, R.; Willcocks, A.L.; Strupish, J.; Nahorski, S.R.; Potter, B.V.L. *Biochem. Soc. Trans.* **1988**, *16*, 992.
(18) Schultz, C.; Metschies, T.; Jastorff, B. *Tetrahedron Lett.* **1988**, *29*, 3919.
(19) Kulagowski, J.J. *Tetrahedron Lett.* **1989**, *30*, 3869.
(20) Falck, J.R.; Abdali, A.; Wittenberger, S.J. *J. Chem. Soc., Chem. Commun.* **1990**, 953-955.
(21) Ho, P.-T.; Kolt, R.J. *Can. J. Chem.* **1982**, *60*, 663.
(22) Engel, R. *Synthesis of Carbon-Phosphorus Bonds*; CRC Press; Boca Raton, FL, 1988; pp 7-21.
(23) Stull, J.T., Univ. Texas Southwestern, personal communication, 1990.
(24) Ely, J.; Catt, K., USPHS National Institutes of Health, personal communication, 1990.
(25) Traynor-Kaplan, A.E.; Harris, A.L.; Thompson, B.L.; Taylor, P.; Sklar, L.A. *Nature* **1988**, *334*, 353-356.
(26) Pignataro, O.P.; Ascoli, M. *J. Biol. Chem.* **1990**, *265*, 1718-1723.
(27) Elliott, J.D.; Hetmanski, M.; Stoodley, R.J.; Palfreyman, M.N. *J.C.S., Perkin 1* **1981**, 1782-1789.
(28) David, S.; Hanessian, S. *Tetrahedron* **1985**, *41*, 643-663.
(29) Dreef, C.E.; Elie, C.J.J.; Hoogerhout, P.; van der Marel, G.A.; van Boom, J.H. *Tetrahedron Lett.* **1988**, *29*, 6513-6516.
(30) Ferguson, M.A.J. *Ann. Rev. Biochem.* **1988**, *57*, 285-320.

RECEIVED February 11, 1991

Chapter 12

Synthesis and Complexation Properties of Inositol–Phosphates

V. I. Shvets[1], A. E. Stepanov[1], L. Schmitt[2], B. Spiess[3], and G. Schlewer[2]

[1]Institute of Fine Chemical Technology, Chair of Biotechnology, Prospekt Vernadskogo Dom 86 117571 Moscow, Union of Soviet Socialist Republics
[2]Centre de Neurochimie du Centre National de la Recherche Scientifique, Département de Pharmacochimie Moléculaire, 5 rue Blaise Pascal, 67084 Strasbourg, France
[3]Faculté de Pharmacie, Laboratoire de Chimie Analytique et de Bromatologie, (IUT–ULP), 74 route du Rhin, 67401 Illkirch, France

Abstract : This chapter deals with synthetic methods of some Inositol-Phosphates (IP) and their complexation properties towards alkali and alkali-earth cations. The most important steps for the syntheses of IPs are : the preparation of selectively protected inositols, the separation of racemates and, the phosphorylation procedures. The methodology involved in all these steps is illustrated with the synthesis of chiral *myo*-inositol-1,4,5-triphosphate (**1**) and related compounds (**16-18**). In a second part, the complexation ability of the IPs is considered. The effect of the ionic environment on the nature and stability of the species in solution is emphasized and the biological significance of the coordination to metals is assessed. Finally, for seven IP_3 the determination of the protonation constants showed the main influence of the position of the phosphate groups on the cyclitol ring.

Our interest in inositol-phosphates (IPs) started twenty years ago for some of us (V.I.S.; A.E.S.) when our investigations turned towards the chemistry and biochemistry of lipids. The phosphatidyl inositols constitute a very important class of compounds in this particular field. The syntheses of such molecules opened, for us, a very large and enthusiastic field of regio- and stereoselective chemistry.

During this time, the biochemical and pharmacological studies of inositol-phosphates, especially D-*myo*-inositol-1,4,5- triphosphate (**1**) grew rapidly and the cell regulator or second messenger function of phosphatidyl-inositols, inositol-phosphate, and diacylglycerol in connection with the endoplasmic reticulum and C-kinase were well demonstrated in numerous types of living cells (*1*).

The inositol-phosphates mechanism of action will ultimately be explained at a molecular level. It is necessary, therefore, to orient the synthesis of pharmacologically active compounds towards the elucidation of the underlying biochemistry and physico-chemistry.

By considering the chemical functions of IPs and their chelating possibilities it appeared to us that the active form of IPs could be determined by the ionic environment of these molecules in the intracellular medium. The ionization state of the phosphate groups, as well as the equilibria between the various complexed

0097–6156/91/0463–0155$06.00/0

species, are governed by both the pH and the ionic content of the cell. In particular, since some IPs lead to calcium mobilization, the interactions with this cation are particularly important. On the other hand, sodium and potassium movements occur throughout the cell membranes wherever phosphoinositides are present. The alkali and alkali-earth cations may also be involved in complexation reactions. Such reactions give rise to species having different charges and in which the spatial position of the phosphate groups may vary significantly from one complex to another. Thus, the knowledge of the nature and the stability of the complexes that form in solution are of prime importance in structure-activity relationships in the field of the inositol-phosphates.

Methodology for the Synthesis of Inositol-Phosphates

In a search for large scale methods of *myo*-inositol phosphate preparation, we began with isolation of the phosphatidylinositols from bovine brain (2). Hydrolysis of these compounds by means of chemical or enzymatic processes gave us some inositol phosphates the structure of which needed chemical confirmation. Apparently, larger amounts of *myo*-inositol phosphates could better be synthesized by non-biological chemical routes.

The three key steps in the syntheses of chiral inositol-phosphates, starting from *myo*-inositol (2), were :
- preparation of selectively protected inositol,
- resolution of racemates into constituent enantiomers,
- phosphorylation procedures.

Syntheses of Selectively Protected *myo*-Inositol. The most commonly used starting material for the synthesis of inositol analogs was (and still is) *myo*-inositol (2). This achiral molecule must be selectively protected before phosphorylation or generation of chiral derivatives. Thus, in our first trials, the *myo*-inositol skeleton was 1,2,4,5,6-pentaacetylated **3** (3) or 1,2,4,5,6-pentabenzylated **4** (4). Such intermediates can be phosphorylated in position 3(1). Adjustment of the reaction conditions gave the opportunity to obtain 1,4,5,6-tetraacetylated **5** (3) or 1,4,5,6-tetrabenzylated **6** (4) derivatives. These types of inositols having the 2- and 3- hydroxyls free, are very interesting intermediates because the relative orientation of the hydroxyls allows for the possibility of regioselective reactions (5).

In our reaction schemes we have also used previously described methods for selective protection such as di-O-cyclohexylidene (6) or di-O-isopropylidene groups (7). In particular we have used 1,2-5,6 di-O-cyclohexylidene *myo*-inositol (7) for optical resolution (8) (Scheme 1).

Resolution into Enantiomers. The second key step for the syntheses of chiral *myo*-inositol-phosphates is the resolution of the enantiomers. To solve this problem we first used chiral derivatives of carbohydrates as resolving agents. Suitably protected inositol racemates were either transesterified (4) or glycosylated (9,10) by means of D-mannose orthoacetate **8** (3,9) or D-glucose orthoacetate **9** (4,10) (Scheme 2). Diastereomeric intermediates were separated either by column chromatography or by selective crystallization.

During our investigations into the synthesis of chiral D-*myo*-inositol-1,4,5-triphosphate (**1**) we also separated 3(1),6(4)-di-O-benzyl-*myo*-inositol-4(6),5-dianilidophosphate (**10**) as diastereomers formed by esterification of position 1 with menthoxyacetic acid (unpublished) (Scheme 3).

Phosphorylation Procedures. The monophosphorylation of protected inositols to yield compounds such as D-*myo*-inositol-1(3)-monophosphate (**11**) (Scheme 4), as well as multiple phosphorylation to give polyphosphates such as D-*myo*-inositol-1,6-diphosphate (**12**), D-*myo*-inositol- 3,4-diphosphate (**13**), D-*myo*-inositol- 1,4-diphosphate (**14**), and D-*myo*-inositol-3,6-diphosphate (**15**), was possible by simply

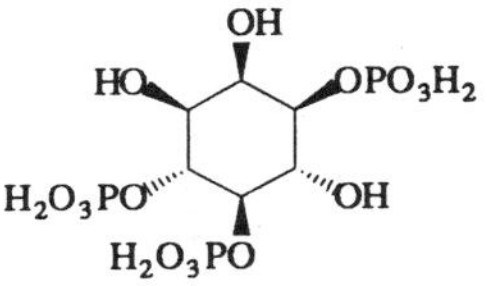

1

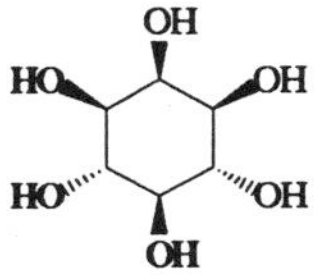

2

Scheme 1 : Selectively Protected *myo*-Inositols

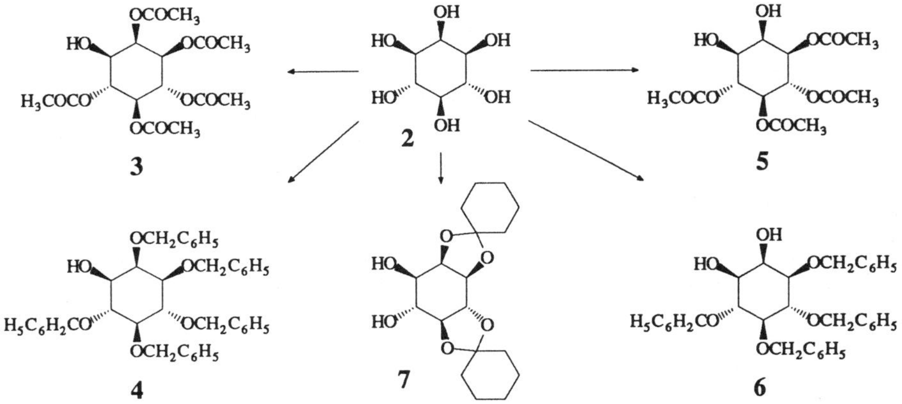

3

2

5

4

7

6

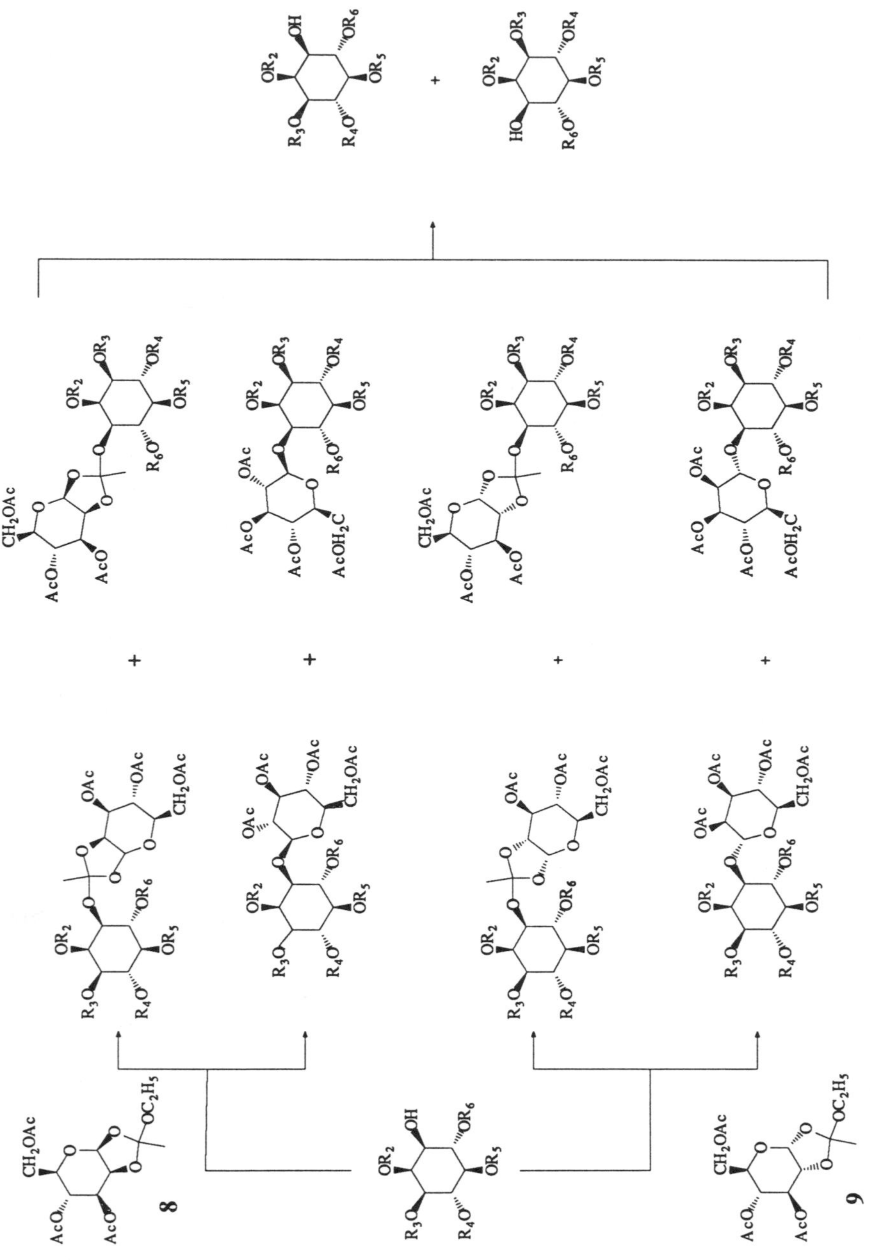

Scheme 2 : Resolution of Enantiomers by Means of Orthoesters or Glycosides

Scheme 3 : Resolution of Enantiomers by Means of Diastereomeric Menthoxyacetic Esters

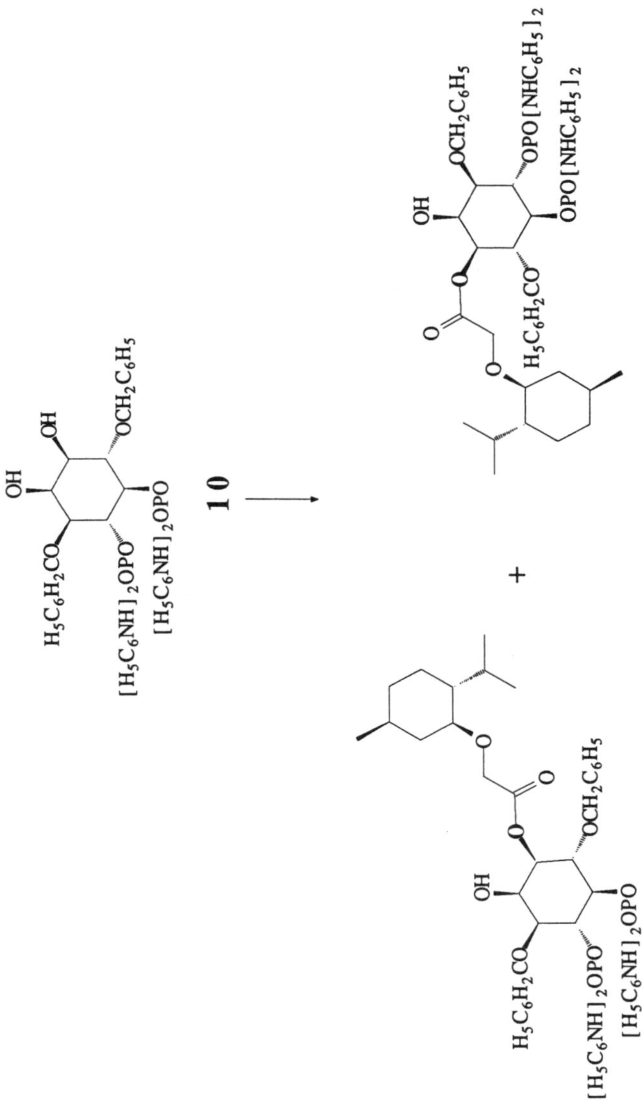

Scheme 4 : Some Chiral *myo*-Inositol Phosphates Prepared from *myo*-Inositol

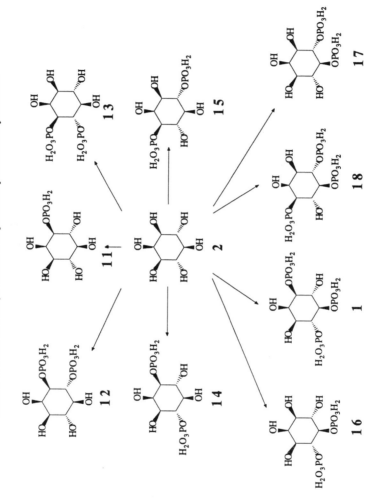

using diphenylphosphoryl chloride *(11)*. Only for the vicinal 4 and 5 positions of the inositol (positions 6 and 5 for the enantiomer) have we had use dianilidophosphoryl chloride to get quantitative phosphorylation **(16-18)** without side reactions such as cyclic phosphate formations *(11,12)*. Unfortunately, though this reagent was efficient for phosphorylation, it seemed not the most suitable owing to the difficulties in removing the anilino protective groups *(12)*.

At this time we prefer to use phosphite *(13)* or pyrophosphate derivatives *(14)* whose protecting groups can be removed under very mild conditions *(13,14)*.

With these methods we were able to prepare the chiral *myo*-inositol phosphates shown in Scheme 4. As an example, the next section will illustrate our approaches to the synthesis of D-*myo*-inositol-1,4,5-triphosphate **(1)**.

Illustration : Total Synthesis of Chiral D-*myo*-Inositol-1,4,5-Triphosphate (1). Our first approach in the synthesis of chiral D-*myo*-inositol-1,4,5-triphosphate **(1)** (and its enantiomer **18**) (Scheme 5) started with the 1,2-O-isopropylidene-3,6-di-O-benzyl-*myo*-inositol **(19)** prepared according to Gigg et al *(15)*. The vicinal diol in positions 4 and 5 was quantitatively phosphorylated by means of dianilidophosphoryl chloride without side reactions to give the diphosphate **20**. The isopropylidene group was then removed to yield the free 1,2-diol **21**.

This compound is our key intermediate. In particular, it is a suitable compound for resolving the racemate. For this resolution we selectively esterified the equatorial hydroxyl group in position 1 by transesterification of D-mannose orthoacetate **8**. The diastereomeric esters were separated by preparative HPLC on silica gel. After hydrolysis of the chiral resolving agent, we obtained the optically active diols **22** and **23**. (the direct total deprotection of these two diols gave D-*myo*-inositol-4,5-diphosphate **(16)** and D-*myo*-inositol-5,6-diphosphate **(17)** respectively *(16)*. The separated diols **22** and **23** can be selectively phosphorylated in the equatorial position 1(3) by means of dianilidophosphoryl chloride, even if in excess. Final deprotection required two steps in a one-pot procedure. The first step removed the anilino groups from the phosphates by means of isoamyl nitrite, which was followed by hydrogenolysis in the presence of palladium on charcoal to give D-*myo*-inositol-1,4,5-triphosphate **(1)** and D-*myo*-inositol-3,5,6-triphosphate **(18)** respectively *(12)*. The final products were stabilized as ammonium salts.

The main drawback of the described synthesis was the final total deprotection. The yields are too low to be applied on a preparative scale. We now prefer the use of phosphites or pyrophosphates as phosphorylating agents. The phosphylation of the vicinal diol **19** with N,N-diethyl-dibenzylphosphoramidite in the presence of tetrazole as catalyst followed, in a one pot procedure, by the oxidation in dibenzylphosphate can be performed with high yield. The protecting groups are chosen (e.g. benzyl groups) so as to allow for simultaneous removal with the other hydroxyl protecting groups by a simple hydrogenolysis *(13,14)*.

Formation and Stability of Inositol-Phosphate Complexes

Almost all the complexation studies published so far have been connected with the well known and easily available *myo*-inositol hexakis(phosphate) or phytic acid. Recently, we have focused our attention on partially phosphorylated derivatives of *myo*-inositol such as IP$_4$, IP$_3$, IP$_2$ and IP$_1$. We report here our results on phytic acid and the various IP$_3$ which were studied.

Phytic Acid. There has long been much interest in this molecule, and several previous studies have dealt with its acid-base behavior *(17-21)*. These studies determined its ionization constants in various supporting-electrolyte backgrounds and ionic strengths. It was surprising to see the discrepancy between the results from one author to another although the experiments were well designed. Moreover, the

Scheme 5 : Synthesis of Chiral 4,5- and 5,6-*myo*-Inositol Diphosphates (16 and 17) and, 1,4,5- and 3,5,6- *myo*-Inositol Triphosphates (1 and 18)

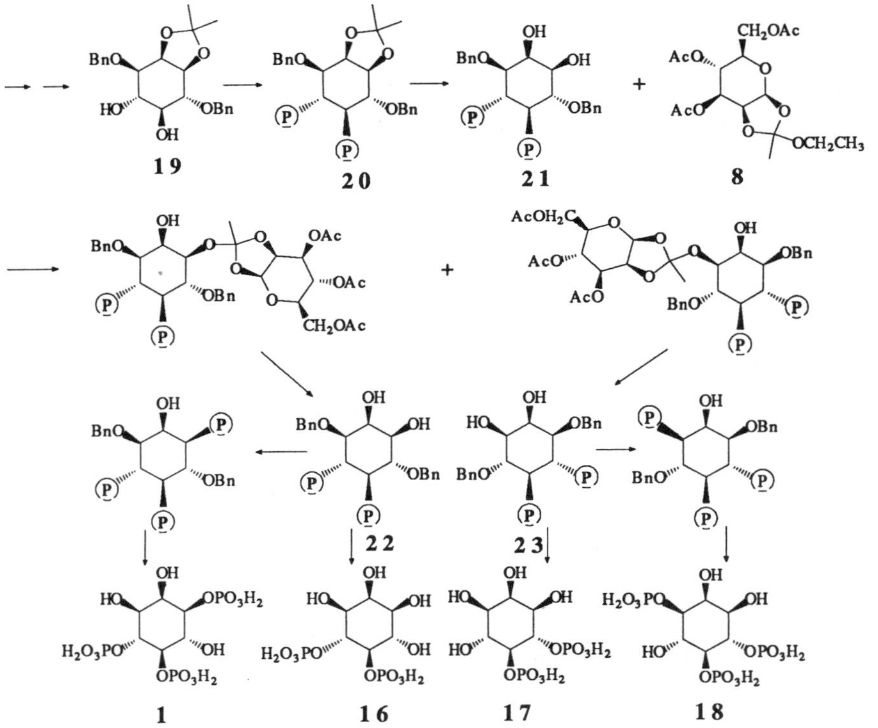

constants that we determined in a n-Bu$_4$NBr medium, strictly in the absence of alkali cations (22), were several orders of magnitude higher than all the constants previously published (17-21). In an attempt to explain these differences, we simulated, in our concentration conditions, the theoretical titration curves resulting from all the data available in the literature. The simulated titration curves and our experimental one are reported in Figure 1. The differences between the curves also show the differences between the constants. It can be observed that the shift of each theoretical curve with our experimental one is dependent on the concentration of supporting-electrolyte containing alkali-metal cations. Such curves are typical of titration curves where complexation occurs, i.e where the interfering cation competes with the proton for the coordination sites. In that case, the shift is especially high, so that the complex is more stable, or for a given complex, that the concentration of the metal increases. A difference of about four pK units for some acidic phosphate groups illustrates the great tendency for phytic acid to form complexes with alkali cations.

In order to gain insight into the complexation properties of phytic acid towards alkali-metal cations, a qualitative approach was undertaken (23). In these studies, the effect of the presence of Li$^+$, Na$^+$, K$^+$ and Cs$^+$ on the protonation constants was considered. The comparison of the protonation constants determined either with (log K$_{yM}$) or without (log K$_y$) metallic cations usually gives valuable information about the extent of the complexation reactions. In addition, this enables the binding properties of phytic acid with different cations of the alkali-metal series to be compared. A suitable way of showing the complexation ability of the ligand is to consider the values of

$$\Delta \log K_{yHM} = \log K_{yH} - \log K_{yM}.$$

These differences are shown in Figure 2 for the protonation steps y=3 to y=6 and for metal to ligand ratios of 11.0 and 5.5, i.e. ratios that remain below those which would result from the dissolution of a dodeca-alkali-metal salt. The results have been extensively discussed mainly in terms of the probable conformation of the complexes (23). Taking into account NMR results obtained by different authors (24,25) and the general shape of the curves, an equatorial prevalent conformation is proposed for pH values up to 9.5. Above this pH, Na$^+$ and K$^+$ seem to stabilize the axial conformer, whereas Li$^+$ and Cs$^+$ promote the equatorial form. The results also confirm, as indicated by the large values of $\Delta \log$ K$_{yHM}$ the strong complexation properties of phytic acid which, in addition, tend to increase as the pH rises (23).

myo-Inositol Triphosphates.

D-*myo*-Inositol-1,2,6-Triphosphate. Among all the inositol triphosphates, the Ins(1,2,6)P$_3$ isomer is the most easily obtained by acidic or enzymatic hydrolysis of phytic acid followed by chromatographic separation. Thus, most of our investigations at first were centered around this compound (26,27). Its protonation and complexation properties towards alkali and alkali-earth cations were studied in two different media and temperature conditions. At 25°C, in a 0.1M n-Bu$_4$NBr medium, the results describe the intrinsic complexation properties of the ligand, whereas at 37°C in a 0.2M KCl (conditions close to those encountered in the cell) the results provide information on the physico-chemical behaviour of the ligand in a biological medium. The nature of the species in equilibrium in both media as well as their stability constants are given in detail in references 26 and 27. The protonation constants determined usually correspond to the first proton of each phosphate group, the second proton being, in general, too acidic to be obtained accurately. The main characteristics of these constants are, on one hand, the large differences between them (in n-Bu$_4$NBr 0.1M, log K$_1$ =9.48, log K$_2$ =7.22, log K$_3$ =5.70) and, on the other hand, the marked difference in regard to simple phosphate monoesters (log K=6.72 for n-BuOPO(OH)$_2$)

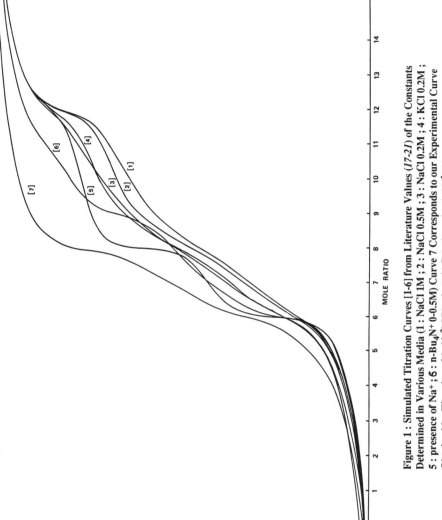

Figure 1 : Simulated Titration Curves [1-6] from Literature Values (17-21) of the Constants Determined in Various Media (1 : NaCl 1M ; 2 : NaCl 0.5M ; 3 : NaCl 0.2M ; 4 : KCl 0.2M ; 5 : presence of Na+ ; 6 : n-Bu₄N+ 0-0.5M) Curve 7 Corresponds to our Experimental Curve Obtained by Titration of 2x10⁻³M Phytic Acid Solution with 1x10⁻¹M base

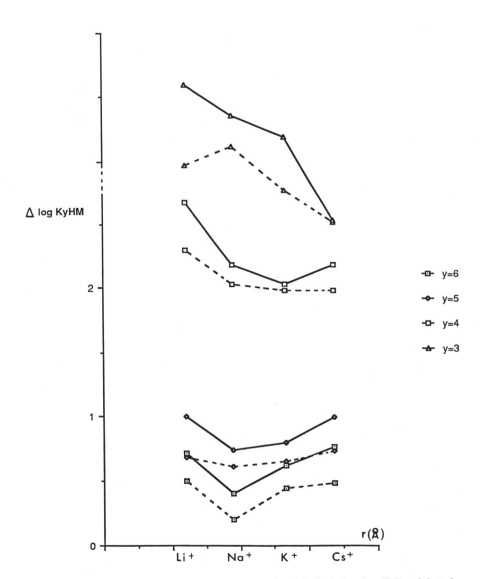

Figure 2 : Values of log K_{yHM} Versus the Ionic Ratios of Cationic Species. (Full and dotted lines correspond to M/IP_6 11.0 and 5.5 respectively)

(28). Such results tend to show that the acidity of the first proton is governed by two opposing effects : one which increases the acidity through the high hydrophilicity of the compound, one which decreases the acidity by the formation of very strong hydrogen bonds. In addition, the differences between the constants obtained in the presence and the absence of K^+ is still a good argument for the complexation ability of the $Ins(1,2,6)P_3$ towards this cation.

A systematic study of the protonation constants in various ionic-strength conditions reported in Figure 3 seems to show that the competition between the potassium and the proton is especially high for potassium concentrations ranging from 0 to 0.2M, i.e. concentrations encountered in biological media.

Besides K^+, other cations such as Li^+, Na^+, Rb^+ and Cs^+ also form complexes of unusually high stability, but no particular selectivity is displayed by $Ins(1,2,6)P_3$ towards these cations (27).

Unlike phytic acid, $Ins(1,2,6)P_3$ forms soluble complexes over a wide pH range with alkali-earth cations. For instance, with Ca^{++} various species including mononuclear, polynuclear and protonated species were found. Their stability constants indicate that such species undoubtedly exist in complex media, especially when the calcium concentration is several times higher than the concentration of the ligand (26).

myo-Inositol-1,4,5-Triphosphate and Other Triphosphates. Recently we extended our investigations to other inositol triphosphates obtained either by synthesis or via separation of mixtures resulting from phytic acid hydrolysis. In particular, much attention was paid to racemic myo-inositol-1,4,5-triphosphate and its acid-base properties. Figure 4 shows the distribution curves calculated for a 0.001M solution of $Ins(1,4,5)P_3$ with the constants that where obtained in the media previously described.

In the 0.1M n-Bu$_4$NBr medium (medium 1), and in the studied pH range (2.5-9.5), equilibria between four protonated species do exist, whereas in a 0.2M KCl medium (medium 2), only three species could be detected. Moreover, in the presence of potassium, the maximum percentage of such species is shifted to lower pH values. This clearly shows, once again, that the ionic environment may contribute to settle the ionic state of the protonated species. In Figure 4 (medium 2), one sees binding data recently published by Worley et al. (29). These authors studied the affinity versus the pH of $Ins(1,4,5)P_3$ with specific receptors in brain membranes. It is remarkable to see that the largest part of the binding curve fits the curve corresponding to the fully deprotonated species very well. The shoulder of the first part of the curve corresponds to the additional binding ability of the monoprotonated species.

Finally, it was interesting to consider the acid-base properties of other IP_3 species, in order to see the influence of the position of the phosphate groups on these properties. Figure 5 displays the three first protonation constants of seven different isomers. The abscissa is labelled according to the position and the conformation of the phosphate groups on the myo-inositol ring (p and q mean equatorial or axial, if p is equatorial, then q is axial and vice versa ; a point refers to a non-phosphorylated position). The different isomers represented are : $Ins(1,5,6)P_3$: (ppp...), $Ins(1,2,3)P_3$: (pqp...), $Ins(1,2,6)P_3$: (ppq...), $Ins(1,4,5)P_3$: (pp.p..) and $Ins(1,3,4)P_3$: (pp.p..), $Ins(2,4,6)P_3$: (p.q.p.), $Ins(1,3,5)P_3$: (p.p.p.). By considering Figure 5 , some important features related to the acid-base properties of the compounds under study can be pointed out. For log K_1 a difference reaching 2.70 pK units is found between the highest and the lowest value. This indicates a marked influence of the position of the phosphate groups. In addition, the basicity of the two phosphates carrying the first protons decreases upon going from vicinal to alternate phosphate positions. For the third phosphate group the reverse effect occurs. Such a tendency is easily understood by the two effects previously mentioned : the hydrophilic effect and the setting of very strong hydrogen bonds favored by a high density of negative charges. In the case of alternate phosphate groups, especially if the conformer is equatorial prevalent, the

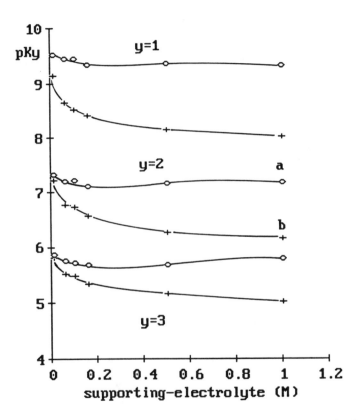

Figure 3 : Curves of the Variation of Stepwise Protonation Constants of Ins(1,2,6)P$_3$ vs Concentration of Supporting Electrolyte (a) : n-Bu$_4$NBr media ; (b) : KCl media

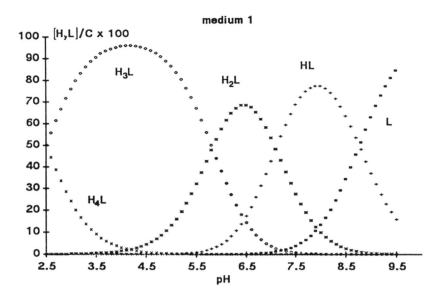

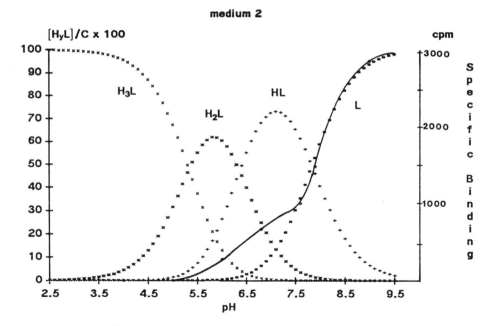

Figure 4 : Distribution Curves of Protonated Species of Ins(1,4,5)P$_3$ Plotted Against pH
(C$_L$ = 0.001M medium 1 : n-Bu$_4$NBr 0.1M, 25°C ; medium 2 : KCl 0.2M, 37°C. The full line
corresponds to the binding data from reference *29*)

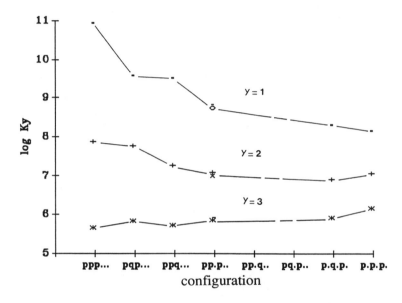

Figure 5 : Variation of the Stepwise Protonation Constants For Various IP₃.

lowest interactions between the phosphates are expected, and the Log K values are then the closest to the log K of a monoesterified phosphate (log K = 6.72 for n-butylphosphate). It is also interesting to note that for the two (pp.p..) isomers, $Ins(1,4,5)P_3$ and $Ins(1,3,4)P_3$, which are diastereomers with inverted configurations at the hydroxyls in position 2 and 6, the log K values only differ slightly, indicating that it is mainly the relative position of the phosphate around the ring that imposes the acid-base properties.

Aknowledgments. We thank the State Commitee for Sciences and Technology, the USSR Committee for People Education, the Ministère des Affaires Etrangères and the Ministère de l'Education Nationale for supporting our France-USSR collaboration (project N° II-A-54) and Robert Kerr for manuscript corrections. The authors are also grateful to PERSTORP PHARMA (Sweden) for providing $Ins(1,5,6)P_3$, $Ins(1,2,6)P_3$, $Ins(1,3,4)P_3$ and to Prof. Akira Hara for providing $Ins(1,2,3)P_3$.

References:

1   Berridge, M. J.; Irvine, R.F. *Nature* **1984**, *312*, 315
2   Klyashchitskii, B. A.; Mezhova, I.V.; Starkova, E.P.; Krasnopol'skii, Y.M.;
    Sennikov, G.A.; Miroshnikov, A.I.: Shvets V.I. *Biotekhnologiya (USSR)*, **1989**, *5*,
    27.
3   Stepanov, A. E.; Tutorskaya, O.O.; Klyashchitskii, B.A.; Shvets, V.I.;
    Evstigneeva, R.P. *Zhur. Obsh. Khim.* **1972**, *42*, 709
4   Shvets, V. I.; Klyashchitskii, B. A.; Stepanov, A. E.; Evstigneeva, R.P.
    Tetrahedron, **1973**, *29*, 331
5   Gigg, R.; Warren, C. D. *J. Chem. Soc. C*, **1969**, 2367
6   Garegg, P. J.; Iversen, T.; Johansson, R.; Lindberg, B. *Carbohydr. Res.,* **1984**, *130*,
    322
7   Gigg, J.; Gigg, R.; Payne, S.; Conant, R. *Carbohydr. Res.,* **1985**, *142*, 132
8   Sadovnikova, M. S.; Kuznetsova, Z. P.; Shvets, V. I.; Evstigneeva, R. P. *Zhur.
    Org. Chim.* **1975**, *11*, 1211
9   Stepanov, A. E.; Shvets, V. I.; Evstigneeva, R. P. *Bioorg. Khim. (USSR),* **1976**, *2*,
    1618
10 Stepanov, A. E. ; Klyashchitskii, B. A.; Shvets, V. I.; Evstigneeva, R. P. *Bioorg.
    Khim (USSR),* **1976**, *2*, 1627
11 Krylova, V. N.; Kobel'kova, N. J.; Oleinik, G. F.; Shvets, V. I. *Zhur. Org. Chim,*
    1980, *16*, 62
12 Stepanov, A. E.; Runova, O. E.; Schlewer, G.; Spiess, B.; Shvets, V. I.
    Tetrahedron Lett., **1989**, *30*, 5125
13 Hamblin, M. R.; Flora, J. S.; Potter, B. V. L. *Biochem. J.,* **1987**, *246*, 771
14 deSolms, S. J.; Vacca, J. P.; Huff, J. R. *Tetrahedron Lett.,* **1987**, *28*, 4503
15 Gigg, J.; Gigg, R.; Payne, S.; Conant, R. *J. Chem. Soc. Perkin Trans. I*, **1987**, 423
16 Stepanov, A.E.; Runova, O.B.; Schlewer, G.; Spiess, B.; Shvets, V.I. *Synthesis of
    Biologically Active Natural Products* ; Proceedings F.E.C.S. Fifth International
    Conference on Chemistry and Biotechnology of Biologically Active Natural
    Products ; Bulgarian Academy of Sciences ; **1989**, Vol. 3 ; 466-470
17 Barre, R.; Courtois, J.E.; Wormser, G.; *Bull. Soc. Chim. Biol.,* **1954**, *36*; 455
18 Costello, A.J.R.; Glonek, T.; Myers, T.C.; *Carbohydr. Res.,* **1976**, *46*, 159
19 Hoff-Jorgensen, E.; *K. Dan Vidensk. Selsk. Nat. Med.,* **1944**, *21(7)*, 1
20 Perrin, D.D.; in *Inositol-Phosphates*, Cosgrove, D.J.; Elsevier, Amsterdam, **1980**
21 Evans, W.J.; McCountney, E.J.; Shrager, R.I., *J. Amer. Oil Chem. Soc.,* **1982**, *59*,
    189
22 Bieth, H.; Spiess, B. *J. Chem. Soc. Faraday Trans I*, **1986**, *82*, 1935
23 Bieth, H.; Jost, P.; Spiess. B.; Wehrer, C.; *Anal. Lett.,* **1989**, *22(3)*, 703

24 Champagne, E.T.; Robinson, J.W.; Gale, R.J.; Nauman, M.A.; Rao, R.M.; Livzzo, J.A., *Anal. Lett.*, **1985** ,*18(19)*, 2471

25 Isbrandt, L.R.; Oertel, R.P., *Anal. Lett.*, **1981**, *102*, 314

26 Bieth, H.; Jost, P.; Spiess, B., *J. Inorg. Biochem.*, **1990**, *39*, 59

27 Bieth, H.; Schlewer, G.; Spiess, B., *J. Inorg. Biochem.,* (in press)

28 Massoud, S.S.; Sigel, H., *Inorg. Chem.*, **1988**, *27*, 1447

29 Worley P.F.; Baraban J.M.; Supattapone S.; Wilson V.S.; Snyder S.H.; *J. Biol. Chem.*, **1987**, *262*, 12132

RECEIVED March 11, 1991

Chapter 13

Phosphorothioate Analogues of Phosphatidylinositol and Inositol 1,2-Cyclic Phosphate

Application to the Mechanism of Phospholipase C

Karol S. Bruzik, Gialih Lin, and Ming-Daw Tsai

Department of Chemistry, The Ohio State University, Columbus, OH 43210

The diastereomers of phosphorothioate analogues of dipalmitoylphosphatidylinositol (DPPsI) have been synthesized from protected optically active *myo*-inositol derivatives. Their configurations at phosphorus have been determined on the basis of stereospecific hydrolysis catalyzed by phospholipase A_2. The reactions catalyzed by phosphatidylinositol-specific phospholipases C (PI-PLC) from *Bacillus cereus* and guinea pig uterus have been shown to be stereospecific toward the R_p isomer of DPPsI. The configuration of one of the products, inositol cyclic-1,2-phosphorothioate (IcPs), was determined from proton and ^{31}P NMR data. The results indicate that the conversion of DPPsI to IcPs, catalyzed by PI-PLC from both sources, proceeds with inversion of configuration at phosphorus, which suggests a direct displacement mechanism.

Phosphorothioate analogues of biophosphates have proven a valuable tool in investigating the mechanisms of action of various phosphohydrolase and phosphotransferase enzymes. Following reports on the synthesis, configurational analysis and biochemical applications of oxygen-labeled and phosphorothioate analogs of nucleotides (*1-5*), we have extended the concept of P-chiral phosphates to phospholipids and used this approach to study the mechanism of phospholipases and other phospholipid-metabolizing enzymes (*6*). This approach has generated two types of information regarding elementary steps in enzymatic reactions: stereospecificity toward one of the two diastereomers of chiral thiophospholipids, and steric course of the reaction involving cleavage of a P-O bond. With the exception of lecithin-cholesterol acyltransferase (*11*), we have found a remarkable stereospecificity of these enzymes toward one of the two diastereomeric thiophospholipids (*6-13*).

The possible fundamental mechanisms of phosphotransfer reactions have been described elsewhere (*14*). The steric course of enzyme-catalyzed phosphotransfer

0097–6156/91/0463–0172$06.00/0

reactions is usually interpreted in terms of the number of steps involving a P-O bond cleavage during the entire catalytic process, since in enzymatic reactions a single nucleophilic displacement at a phosphorus atom usually proceeds with inversion of configuration (*1-14*). Thus an overall inversion of configuration at phosphorus suggests that the reaction proceeds by a single-step mechanism involving an in-line arrangement of the incoming nucleophile and the departing leaving group in the transition state. In very unusual situations the overall inversion could also result from a three-step mechanism. An overall retention of configuration suggests formation of an enzyme-substrate intermediate. In such cases two consecutive steps, each occurring by a single-displacement mechanism (with an inversion of configuration at phosphorus), are necessary to convert a substrate into a product.

The stereochemical information of phospholipases is of particular significance since it is usually difficult to perform detailed kinetic studies due to the surface specificities of these enzymes. The crucial step in determining the steric course of the enzymatic reaction is the configurational assignment of both the substrate and the product. In principle, determination of configurations of P-chiral phosphorothioate analogues of phospholipids can be made on the basis of the above mentioned stereoselectivity of phospholipases A_2 and C with different substrates. For a more comprehensive discussion on the stereochemistry of phospholipases the reader is referred to a recent review (*6*).

Phosphatidylinositides-specific phospholipase C (PI-PLC, E.C.3.1.4.10) is an enzyme catalyzing hydrolysis of phosphatidylinositol (PI, **1**) (or its phosphorylated derivatives) to produce three products: *myo*-inositol-1-phosphate (IP, **2**), *myo*-inositol cyclic-1-phosphate (IcP, **3**), and diacylglycerol (*15,16*) (Scheme 1). This is a critical step in the transduction of the extracellular hormone stimulus into an intracellular physiological signal (*17-19*). The PI-PLC from bacterial and mammalian sources differ greatly in their protein sizes, amino acid sequences and metal-ion requirements, and in the products of their reactions (*15,16,20,21*). Mammalian enzymes usually require Ca^{2+} for their activity and produce a mixture of IcP and IP, whereas *Bacillus cereus* PI-PLC is metal ion independent and forms only IcP. For mammalian enzymes the ratio IcP/IP varies somewhat depending on the enzyme source and reaction conditions.

This chapter describes synthesis and configurational assignment of the phosphorothioate analogues of DPPI (dipalmitoyl-PI) and IcP (abbreviated as DPPsI and IcPs, respectively) and the results of the stereochemistry of the reactions catalyzed by PI-PLC from *B. cereus* and two isozymes (I and II) from guinea pig uterus (*22-24*).

Synthesis of DPPsI. Three methods have been developed to synthesize DPPsI. In the first method (Scheme 2) D-2,3:5,6-di-*O*-cyclohexylidene-4-*O*-methoxymethyl-*myo*-inositol (**4**) was phosphitylated with chloro-(*N*,*N*-diisopropylamino)methoxyphosphine (**5**) (*25*) in the presence of triethylamine in dichloromethane (*22,23*). The resulting phosphoramidate **6** was subjected to the reaction with 1,2-dipalmitoyl-*sn*-glycerol under tetrazole catalysis in THF-acetonitrile, and subsequently the sulfur addition. The mixture of the diastereomeric phosphorothionates (**7a,b**) could not be resolved by chromatographic techniques and was further subjected to deprotection steps with 80% aqueous acetic acid (to remove

Scheme 1

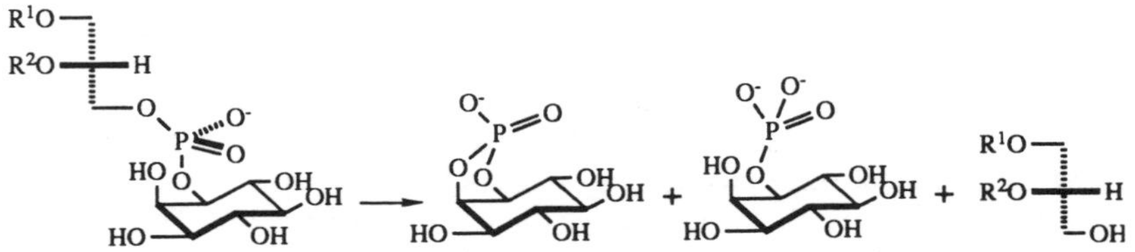

PI, 1   IP, 2   IcP, 3

R^1 = stearoyl; R^2 = arachidonoyl

Scheme 2

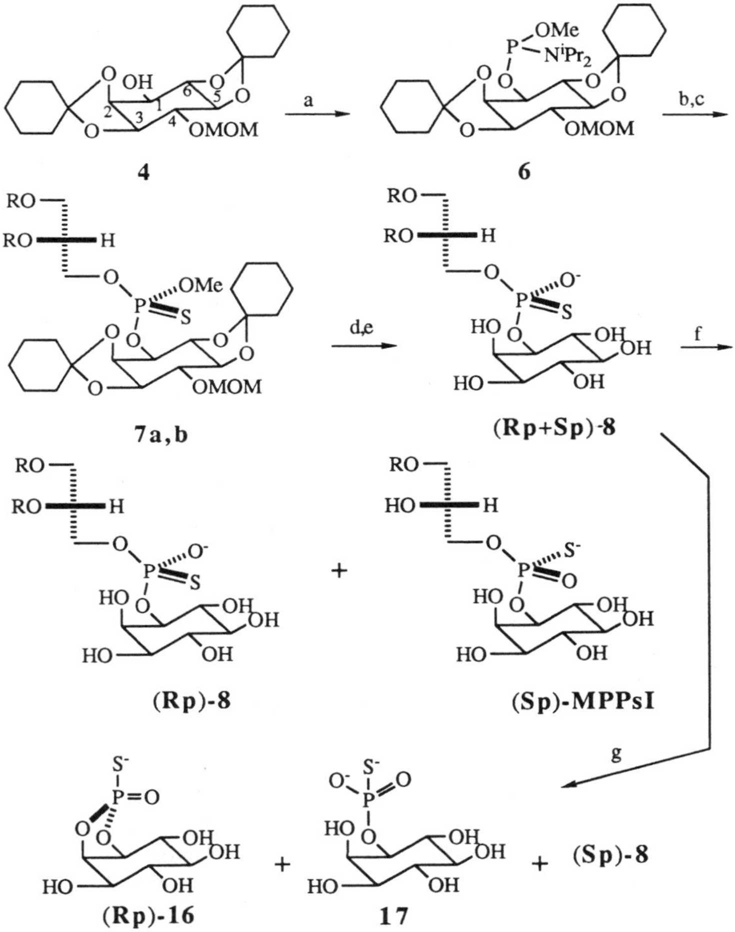

a) ClP(OMe)N^{i}Pr$_2$ (**5**), Et$_3$N; b) 1,2-dipalmitoyl-*sn*-glycerol, tetrazole; c) S$_8$;
d) 80% AcOH; e) NMe$_3$, f) phospholipase A$_2$; g) PI-PLC; R = palmitoyl

acetal protective groups), followed with trimethylamine in toluene (for demethylation). The DPPsI **8** thus synthesized is a mixture of two diastereomers and gives two signals of equal intensity at 57.45 and 57.05 ppm in a ^{31}P NMR spectrum. Separation of diastereomers was effected enzymatically by taking advantage of the stereospecificity of phospholipase A$_2$ (PLA2) and PI-PLC, as described in the next section.

Both the second and the third methods give separate isomers of DPPsI through chromatographic separation of one of the intermediates. The second method (Scheme 3) employed the same phosphitylation steps as described in method 1, but used 1,2-dipalmitoyl-*sn*-glycerol as the starting material and L-1,2,4,5,6-pentabenzyl-*myo*-inositol (**9**) as the second hydroxyl component for the synthesis (*24*). The mixture of R_p and S_p triesters **10a,b** (at the ratio ca. 58:42) could be separated by column chromatography on silica gel using carbon tetrachloride-acetone (40:1) as an eluting solvent. Separated diastereomers **10a** and **10b** were demethylated with trimethylamine and debenzylated with BF$_3$-etherate/ethanethiol to produce DPPsI **8a** and **8b**, respectively.

The third method (Scheme 4) has been developed recently (Bruzik, K. S., Tsai, M.-D., unpublished results) and employs D-camphor-protected inositol derivative **11**, which was synthesized from *myo*-inositol and D-camphor dimethyl acetal in an one-pot procedure as described elsewhere (*26*). Treatment of **11** with *tert*-butyldiphenylsilyl (TBDPS) chloride resulted in a remarkable selective protection of the 1-hydroxyl group, and the resulting triol **12** was fully blocked with methoxymethylene (MOM) groups to give **13**. Deprotection by tetrabutylammonium fluoride yielded **14** as the starting substrate for phosphitylation. The synthesis of phosphorothionates **15a,b** was carried out in an analogous way as described above. The diastereomers **15a** and **15b** could be separated by silica gel chromatography (hexane-acetone/20:1), and were then subjected to demethylation with trimethylamine followed by acid-catalyzed cleavage of all acetal protective groups to give **8a** and **8b**, respectively.

All three procedures employ similar phosphorylating agents, and the yields are all in the range of 40-60% from the protected D-*myo*-inositol. The more difficult part of the synthesis is really in the synthesis of the protected D-*myo*-inositol (**4**, **9**, and **14** in methods 1, 2, and 3, respectively). The syntheses of **4** and **9** have not been shown in the schemes, but they are significantly lengthier than the synthesis of **14** and also involve camphor (method 2) or camphanic acid (method 1) derivatives in the separation of protected D- and L-*myo*-inositols. Method 3 offers a significant short-cut due to the one-pot synthesis of **11** in enantiomerically pure form and the remarkable selectivity of the derivatization of **11** with the sterically bulky *tert*-butyldiphenylsilyl function. Although the synthetic work reported here was primarily aimed at the synthesis of DPPsI, the improvements of the synthetic procedures can also be applied to the synthesis of natural PI and IP.

Configurational Assignment of DPPsI. It has been established previously that PLA2 from various sources hydrolyze specifically the R_p isomers of the phosphorothioate analogues of phosphatidylcholine (DPPsC) and phosphatidylethanolamine (DPPsE) (*6, 7, 10, 27*). Since the phosphorus stereospecificity of PLA2 depends on the stereospecific interaction involving the

Scheme 3

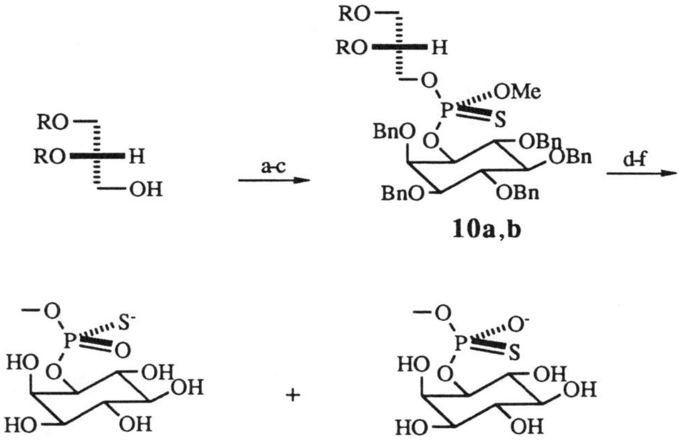

10a,b

8b (Sp)   +   8a (Rp)

a) **5**, Et$_3$N; b) L-1,2,4,5,6-pentabenzyl-*myo*-inositol (**9**), tetrazole; c) S$_8$; d) separation; e) NMe$_3$; f) BF$_3$·Et$_2$O-EtSH

Scheme 4

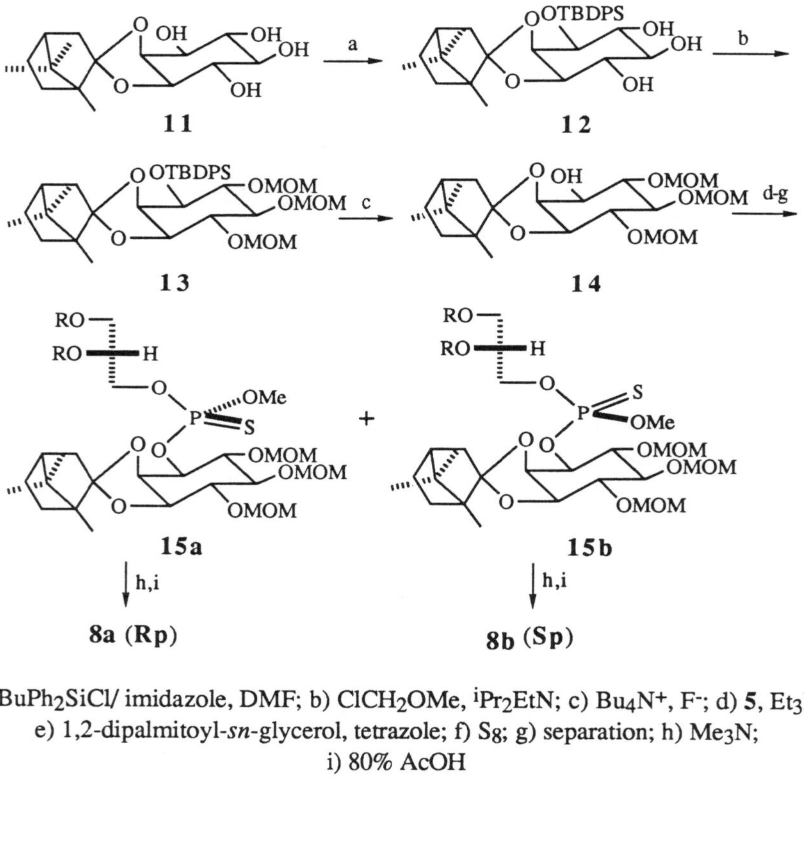

a) ^{t}BuPh$_2$SiCl/ imidazole, DMF; b) ClCH$_2$OMe, ^{i}Pr$_2$EtN; c) Bu$_4$N$^+$, F$^-$; d) **5**, Et$_3$N;
e) 1,2-dipalmitoyl-sn-glycerol, tetrazole; f) S$_8$; g) separation; h) Me$_3$N;
i) 80% AcOH

phosphate group with Ca^{2+} (*10*) and possibly other residues at the active site of the enzyme, and since PLA2 is relatively nonspecific to the structure of the head group of phospholipids, the DPPsI isomer of the same chirality should be the preferred substrate for PLA2. However, due to a reversal of the relative priority of phosphorus substituents in DPPsI relative to DPPsC or DPPsE, R_p and S_p isomers of DPPsI correspond to S_p and R_p isomers, respectively, of DPPsC and DPPsE, in terms of the stereochemical structure of the phosphorothioate group.

As shown in Figure 1, treatment of (R_p+S_p)-DPPsI (spectrum a) in its micellar form with PLA2 resulted in the hydrolysis of the isomer corresponding to the lower field signal in ^{31}P NMR as illustrated by the gradual decrease in its intensity (spectra b-d). The chemical shift of the product (S_p)-MPPsI (Scheme 2, MP stands for 1-monopalmitoyl) nearly coincides with that of the unhydrolyzed (R_p)-DPPsI (isomer of the higher field signal), but they can be resolved in the expanded spectrum e. It was thus concluded that the configurations of isomers giving rise to the signals at δ 57.05 and 57.45 ppm (**8a** and **8b**, respectively) are R_p and S_p, respectively.

(R_p+S_p)-DPPsI as a Substrate of PI-PLC. In the case of *B. cereus* PI-PLC the reaction with (R_p+S_p)-DPPsI as a substrate leads to the formation of inositol cyclic-1,2-phosphorothioate (IcPs) (**16**) (Scheme 2), which gives rise to a ^{31}P NMR signal at 71.4 ppm (Figure 2, spectrum b). At the same time the resonance of the R_p isomer decreases. Both isozymes I and II of PI-PLC from guinea pig uterus are also specific to the R_p isomer of DPPsI and produce the same cyclic product **16**. In addition, formation of the acyclic *myo*-inositol 1-phosphorothioate (**17**, IPs) was observed at the ratio IcPs/IPs ca. 2 for isozyme I and 0.5 for isozyme II. These values were close to IcP/IP ratios observed for natural PI as substrate (1 and 0.5 for isozymes I and II, respectively). These results indicate that all three PI-PLC show the same stereospecificity (prefer the R_p isomer of DPPsI) as the non-PI-specific PLC (prefers the S_p isomers of DPPsC and DPPsE) (*6*). In order to elucidate the steric course of the formation of the IcPs it is necessary to determine the phosphorus configuration of the product **16**.

Synthesis and Configurational Analysis of *cis* and *trans* IcPs. In order to assign the configuration of **16** both isomers of IcPs were synthesized independently as outlined in Scheme 5, starting from DL-1,4,5,6-tetrabenzyl-*myo*-inositol (**18**). Treatment of **18** with **5** in the presence of diisopropylethylamine and subsequently with tetrazole resulted in the formation of a mixture of diastereomeric cyclic phosphites **19** to which elemental sulfur was added in the next step. The resulting phosphorothionates **20** were separated by chromatography and deprotected with lithium in THF-ammonia at -78° to give two diastereomers **16a** and **16b** resonating at δ 71.4 and 70.5 ppm, respectively, in ^{31}P NMR. The configurational assignment of IcPs isomers was made indirectly based on the following findings:

a) The yields of **16a** and its precursors **20a** and **19a** are much lower than those of **16b**, **20b** and **19b**, respectively. On the basis of related studies of a series of 4,5-disubstituted 1,3,2-dioxaphospholanes (*28*), the more stable and predominant isomer **19b** should have the *trans* geometry as indicated in Scheme 5. The oxidation and the

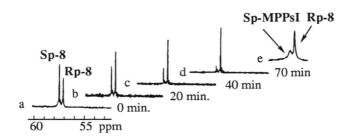

Figure 1. ^{31}P NMR spectra showing a time course of the hydrolysis of (R_p+S_p) DPPsI (5 mM, pH 7.2, 4% Triton X-100) in the presence of phospholipase A_2 (bee venom) at 310 K; spectrum e shows an expanded upfield signal from spectrum d. (Adapted with revision from reference 24)

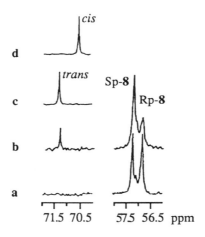

Figure 2. ^{31}P NMR spectra showing stereospecificity of PI-PLC (*B. cereus*) catalyzed hydrolysis of (R_p+S_p)-DPPsI (sample conditions analogous to Figure 1). a) DPPsI prior to adding PI-PLC; b) 72 h after adding 2 mg of PI-PLC; c) *trans*-IcPs (**16a**); d) *cis*-IcPs (**16b**). (Adapted with revision from reference 22)

Scheme 5

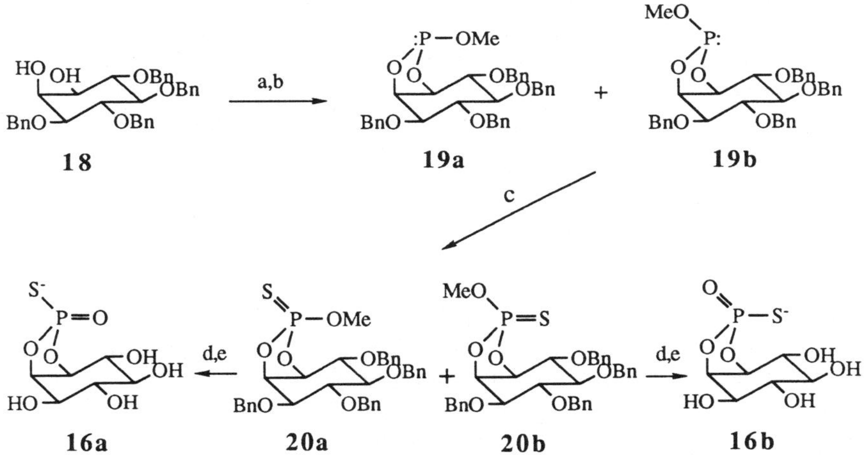

a) **5**, ^{i}Pr$_2$EtN; b) tetrazole; c) S$_8$; d) separation; e) Li/THF-NH$_3$

sulfur addition to cyclic P(III) compounds (such as **19**) usually proceeds with a retention of configuration, and the isomeric ratio is generally preserved (*28*). Thus **20a** and **20b** should be *trans* and *cis*, respectively. The *cis/trans* notations are based on the geometric relationship between the inositol ring and the other larger substituent of phosphorus (which is the OMe group in the case of **19**, but the sulfur atom in the case of **20**).

b) Comparison of model compounds (*28*) suggest that the ^{31}P chemical shift of the *cis* isomer **20b** should be 0.5-2.5 ppm more upfield than that of the corresponding *trans*-isomer **20a**. The observed chemical shifts **20a** and **20b** are 85.9 and 84.2 ppm, respectively.

c) The ^{1}H-^{31}P vicinal coupling constants (between P and H-1 of the inositol ring) of 18.4 and 9.7 Hz for **20a** and **20b**, respectively, correlate well with the values calculated from the Karplus equation (*29*) (18.3 and 9.8 Hz, respectively) for MM-2 optimized conformations of these isomers.

d) The ^{1}H 2D-NOESY spectra show a through-space interactions between the *O*-methyl group and H-4 of inositol in **20a** but not in **20b**. On the other hand, interactions between *O*-methyl group and H-2 are observed in **20b** but not in **20a** (*23*).

We therefore conclude that **20a** has a *trans* and **20b** a *cis* geometry. The same are true for isomers **16a** and **16b**, respectively, since the cleavage of methyl ester in **20** does not involve bond breaking around phosphorus. The absolute configuration at phosphorus of **20a** and **16a** as shown in Scheme 5 should be R_p according to the Cahn-Ingold-Prelog rules (30). Since the synthesis of IcPs starts with DL-**18**, **20a** and **16a** are actually racemic mixtures (D-R_p+L-S_p), so are **20b** and **16b** (D-S_p+L-R_p).

Steric Course of the Formation of IcP Catalyzed by PI-PLC. The results presented above establish that the IcPs produced from (R_p)-DPPsI is the *trans* isomer **16a**. This statement is a corollary to the conclusion that *PI-PLC catalyzed conversion of PI to IcP proceeds with an inversion of configuration* at phosphorus, most likely via direct, in-line attack at phosphorus by the 2-hydroxyl group of the inositol ring.

Implications on the Mechanism of PI-PLC. Figure 3 outlines six possible mechanisms of mammalian PI-PLC. Mechanisms E and F can be excluded because no incorporation of ^{18}O label into IcP was found when the reaction was carried out in [^{18}O]water (*31, 32*). Mechanisms B and D involve an enzyme-phosphoinositol intermediate and predict retention of configuration for the formation of IcP or IcPs. Thus mechanisms B and D can be ruled out based on the observed inversion of configuration at phosphorus. The two remaining possible mechanisms are A and C. Mechanism C resembles the mechanism of ribonuclease A (*33*) in that a cyclic intermediate is formed as the precursor of the hydrolytic product, but also differs in the sense that the cyclic intermediate is not released from ribonuclease A under normal catalytic conditions. Mechanism A represents a unique mechanism in that two reaction pathways proceed in parallel from a common substrate.

Mechanism A is the favored mechanism for mammalian PI-PLC on the basis of the following evidence: (i) there is no detectable conversion of IcP to IP catalyzed by the enzyme (*15, 20, 34, 35*); (ii) mechanism C would predict an increase of the proportion

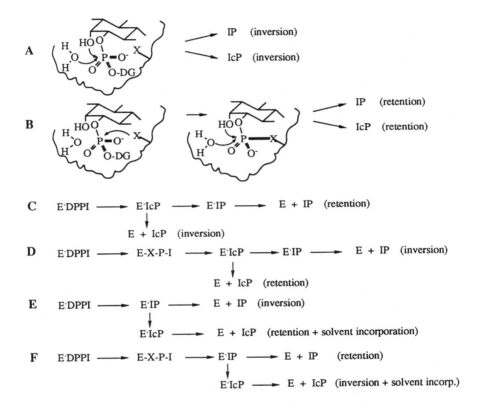

Figure 3. Possible mechanisms of mammalian PI-PLC: A) parallel reactions, direct displacement; B) parallel reactions, covalent enzyme-substrate intermediate; C) consecutive reactions (IcP → IP), direct displacement; D) consecutive reactions (IcP → IP), covalent enzyme-substrate intermediate; E) consecutive reactions (IP → IcP), direct displacement; F) consecutive reactions (IP → IcP), covalent enzyme-substrate intermediate. DG stands for the diacylglyceryl moiety of DPPsI. (Adapted with revision from reference 23)

of the cyclic product in the case of the phosphorothioate substrate since sulfur substitution would most likely slow down the enzyme-mediated hydrolysis of IcPs, but we have observed that the ratio IPs/IcPs is comparable to the ratio IP/IcP, as mentioned in a previous section. However, it should be noted that neither evidence is absolute. The kinetic competence of an intermediate in enzyme catalysis depends on specific rate constants and equilibrium constants (36). The ratio IPs/IcPs could also be fortuitously similar to the ratio IP/IcP in mechanism C.

Recently PI-PLC from *B. cereus* has been shown to exhibit a thousand-fold lower cyclic phosphodiesterase activity (i.e., conversion of IcP to IP) in addition to the main phosphotransferase activity (37). This observation strongly suggests that mechanism C is the actual mechanism for PI-PLC from *B. cereus*. Since the specific activity of mammalian PI-PLC is usually thousand-fold lower than that of *B. cereus* PI-PLC, it is possible that mammalian and bacterial enzymes both utilize mechanism C, but bacterial enzymes have evolved to greatly improve the first step and the release of the cyclic intermediate. Alternatively, mammalian and bacterial enzymes could have evolved to different mechanisms (parallel and sequential pathways, respectively). There is at least one closely related precedent: phosphatidylserine synthase from yeast catalyzes the reaction via a sequential mechanism, whereas that from *Escherichia coli* uses a ping-pong mechanism (13). In both cases one of the differences between the enzymes from lower and higher organisms is requirement for divalent metal ions in the latter. The unequivocal distinction between mechanisms A and C for mammalian PI-PLC is awaiting elucidation of the steric course of the formation of IP from PI.

Acknowledgments. This work was supported by Grant GM 30327 (to M.-D. T.) from National Institutes of Health and Grant CPBP.01.13.3.16 (to K. S. B.) from Polish Academy of Sciences. We thank Dr. C. F. Bennett for providing us the mammalian PI-PLC isozymes used for the work described in this chapter. This paper is part 24 in the series "Phospholipids Chiral at Phosphorus". For part 23, see reference 38.

References

1. Eckstein, F. *Angew. Chem. Int. Ed. Engl.* **1983**, *22*, 423.
2. Frey, P. A. *Tetrahedron* **1982**, *38*, 1541.
3. Gerlt , J. A., Coderre, J. A., Mehdi, S. *Adv. Enzymol. Relat. Areas Mol. Biol.* **1983**, *55*, 291.
4. Buchwald, S. L., Hansen, D. E., Hassett , A., Knowles, J. R., *Methods Enzymol.* **1982**, *87*, 279.
5. Lowe, G. *Acc. Chem. Res.* **1983**, *16*, 244.
6. Bruzik, K. S., Tsai, M.-D. *Methods Enzymol.* **1991** *197*, 258.
7. Bruzik, K., Jiang, R.-T., Tsai, M.-D. *Biochemistry* **1983**, *22*, 2478.
8. Bruzik, K., Tsai, M.-D. *Biochemistry* **1984**, *23*, 1656.
9. Jiang, R.-T., Shyy, Y.-J., Tsai, M.-D. *Biochemistry* **1984**, *23*, 1661.
10. Tsai, T.-C., Hart, J., Jiang, R.-T., Bruzik, K., Tsai, M.-D. *Biochemistry* **1985**, *24*, 3180.

11. Rosario-Jansen, T., Pownall, H. J., Jiang, R -T., Tsai, M-D. *Bioorg. Chem.* **1990**, *18*, 179.

12. Rosario-Jansen, T., Jiang, R.-T., Tsai, M.-D., Hanahan, D. J. *Biochemistry* **1988**, *27*, 4619.

13. Raetz, C. R. H., Carman, G. M., Dowhan, W., Jiang, R.-T., Waszkuc, W., Loffredo, W., Tsai, M.-D. *Biochemistry* **1987**, *26*, 4022.

14. Knowles, J. R. *Annu. Rev. Biochem.* **1980**, *49*, 877.

15. Majerus, P. W., Connolly, T. M., Deckmyn, H., Ross, T. S., Bross, T. E., Ishii, H., Bansal, V. S., Wilson, D. B. *Science* **1986**, *234*, 1519.

16. Rhee, S. G., Suh, P.-G., Ryu, S.-H., Lee, S. Y. *Science* **1989**, *244*, 546.

17. Berridge, M. J., Irvine, R. F. *Nature* **1984**, *312*, 315.

18. Berridge, M. J. *Annu. Rev. Biochem.* **1987**, *56*, 159.

19. Berridge, M. J., Irvine, R. F. *Nature* **1989**, *341*, 197.

20. Dawson, R. C. M., Freinkel, N., Jungalwala, F. B., Clarke, N. *Biochem. J.* **1971**, *122*, 605.

21. Kim, J. W., Ryu, S. H., Rhee, S. G. *Biochem. Biophys. Res. Commun.* **1989**, *163*, 177.

22. Lin, G., Tsai, M -D. *J. Am. Chem. Soc.* **1989**, *111*, 3099.

23. Lin, G., Bennett, C. F., Tsai, M.-D. *Biochemistry* **1990**, *29*, 2747. When comparing the results of this reference and the present chapter, it should be noted that the *exo* and *endo* notations in ref. 23 correspond to the *trans* and *cis* notations, respectively, in this chapter, and that all ^{31}P chemical shifts in ref. 23 should be corrected by +1.52 ppm.

24. Salamonczyk, G. M., Bruzik, K. S. *Tetrahedron Lett..* **1990**, 2015.

25. Bruzik, K. S., Salamonczyk, G., Stec, W. J., *J. Org. Chem.* **1986**, *51*, 2368.

26. Bruzik, K. S., Salamonczyk, G. M. *Carbohydr. Res.* **1989**, *195*, 67.

27. Orr, G. A., Brewer, C. F., Heney, G. *Biochemistry*, **1982**, *21*, 3202.

28. Mikolajczyk, M., Witczak, M. *J. Chem. Soc. Perkin Trans 1* **1977**, 2213.

29. Bentrude, W. G., Setzer, W. N. in *Phosphorus-31 NMR Spectroscopy in Stereochemical Analysis*, Verkade, J. G., Quin, L. D., Eds. VCH Publishers, Deerfield Beach, 1987, pp 365-389.

30. Cahn, R. S., Ingold, C. K., Prelog, V. *Angew. Chem. Int. Ed. Engl.* **1966**, *5*, 385.

31. Wilson, D. B., Bross, T. E., Hofmann, S. L., Majerus, P. W. *J. Biol. Chem.* **1984**, *259*, 11718.

32. Wilson, D. B., Bross, T. E., Sherman, W. R., Berger, R. A., Majerus, P. W. *Proc. Natl. Acad. Sci. U. S. A.* **1985**, *82*, 4013.

33. Eckstein, F. *Angew. Chem. Int. Ed. Engl.* **1975**, *14*, 160.

34. Allan, D., Michell, R. H. *Biochem. J.* **1974**, *142*, 591.

35. Lapetina, E. G., Michell, R. H. *Biochem. J.* **1973**, *131*, 433.

36. Cleland, W. W. *Biochemistry* **1990**, *29*, 3194.

37. Volwerk, J. J., Shashidhar, M. S., Kuppe, A., Griffiths, O. H. *Biochemistry* **1990**, *29*, 8056.

38. Loffredo, W. M., Jiang, R.-T., Tsai, M.-D. *Biochemistry* **1990**, *29*, 10912.

RECEIVED February 11, 1991

Chapter 14

Phosphorothioate Analogues of D-*myo*-Inositol 1,4,5-Trisphosphate

Chemistry and Biology

Barry V. L. Potter

School of Pharmacy and Pharmacology, University of Bath, Claverton Down, Bath BA2 7AY, England

D-*myo*-inositol 1,4,5-trisphosphate [Ins-(1,4,5)P_3] is a second messenger that mediates mobilization of intracellular Ca^{2+}. Chemical modification of Ins(1,4,5)P_3 can yield analogues with novel properties towards the Ca^{2+}-mobilizing receptor and the metabolic enzymes Ins(1,4,5)P_3 5-phosphatase and 3-kinase. We have synthesized phosphorothioate analogues of Ins(1,4,5)P_3 which are finding many biological applications as potent non-hydrolysable agonists and enzyme inhibitors. The chemical synthesis and biological properties of these compounds are reviewed.

D-*myo*-inositol 1,4,5-trisphosphate [Ins(1,4,5)P_3 (1), Figure 1]* is a second messenger coupling agonist stimulation of cell surface receptors to the release of intracellular Ca^{2+} (*1*). Because of the fundamental importance of the polyphosphoinositide signalling system in cell biology it is essential to have access to chemical tools which can facilitate pharmacological intervention at Ins(1,4,5)P_3 receptors and the metabolic enzymes which act upon this second messenger (*2*). Indeed, Ins(1,4,5)P_3 antagonists and compounds which block signal transduction by the polyphosphoinositide pathway may have therapeutic value as potential drugs (*3*), provided they can gain internal access to

* Unless otherwise indicated, 'inositol' refers to the *myo*-isomer

0097–6156/91/0463–0186$06.00/0

cells. Heparin has been found to act as an $Ins(1,4,5)P_3$ antagonist (4), but few such tools have yet been identified and there are significant difficulties intrinsic to a drug design strategy based upon $Ins(1,4,5)P_3$.

Now that problems with inositol phosphate synthesis have essentially been overcome, it is possible to envisage the rational design and chemical synthesis of $Ins(1,4,5)P_3$ analogues (5-7). Few structurally-modified compounds possessing biological activity have, however, yet been prepared. The first example of such a compound was *myo*-inositol 1,4,5-trisphosphorothioate [$Ins(1,4,5)PS_3$ (2), Figure 1] (8). Other phosphorothioate analogues of $Ins(1,4,5)P_3$ have now been synthesized i.e. inositol 1,4-bisphosphate-5-phosphorothioate [$Ins(1,4,5)P_3$-5S, (3)] (9), inositol 1-phosphorothioate-4,5-bisphosphate [$Ins(1,4,5)P_3$-1S, (4)] (10) and inositol 1-phosphate-4,5-bisphosphorothioate [$Ins(1,4,5)P_3$-4,5S, (5)] (N. J. Noble & B. V. L. Potter, unpublished). Extensive studies have demonstrated that the unique properties of inositol phosphorothioates, especially their resistance to degradation by intracellular phosphatases (11,12), will make them valuable tools for investigation of the actions and metabolism of inositol phosphates. This chapter will address only inositol phosphorothioates. For details of other inositol and inositol phosphate analogues which have been synthesized the reader is directed to recent reviews (5-7).

Synthesis

Many synthetic routes to the natural second messenger, $Ins(1,4,5)P_3$ have now been devised (5-7), but will not be discussed here. Our initial route (8) was based upon P(III) methodology, used successfully to prepare *myo*-inositol 4,5-bisphosphate (13) and involved phosphitylation of the protected precursor 1,2,4-tri-*O*-benzyl-*myo*-inositol (14) [Figure 2]. Conversion of the resulting trisphosphoramidite to the hexacyanoethyl trisphosphite was followed by oxidation to the fully protected trisphosphate triester and removal of all protecting groups in one step using sodium in liquid ammonia. The resulting $Ins(1,4,5)P_3$, which was purified by ion-exchange chromatography, was fully active biologically at mobilising intracellular Ca^{2+} and binding to the cerebellar $Ins(1,4,5)P_3$ receptor. The phosphorothioate analogue $Ins(1,4,5)PS_3$ could be prepared by modification of this route by oxidation of the trisphosphite hexaester with a solution of sulfur in pyridine followed by deprotection and purification as for $Ins(1,4,5)P_3$.

Once the biological activity of $Ins(1,4,5)PS_3$ had been established, the potential importance of similar

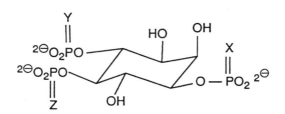

$$X = Y = Z = O \quad (1)$$

$$X = Y = Z = S \quad (2)$$

$$X = Y = O; Z = S \quad (3)$$

$$Y = Z = O; X = S \quad (4)$$

$$Y = Z = S; X = O \quad (5)$$

Figure 1. Structures of Ins(1,4,5)P$_3$ and phosphorothioate analogues. D-isomers are shown.

analogues possessing only single phosphorothioate substitution was realized. In particular, synthesis of the analogue Ins(1,4,5)P$_3$-5S (3) possessing only a 5-phosphorothioate group seemed desirable, since this molecule would be nearer in structure to Ins(1,4,5)P$_3$ than Ins(1,4,5)PS$_3$ and yet would enjoy the advantages of 5-phosphatase resistance. Our synthetic route to this molecule was developed using a combination of P(III) and P(V) chemistry (9) [Figure 3]. Thus, phosphorylation of 2,3,6-tri-*O*-benzyl-5,6-*O*-isopropylidene-*myo*-inositol with bis(2,2,2-trichloroethyl) phosphorochloridate followed by removal of the isopropylidene group and careful phosphorylation of the resulting diol gave a mixture of the 1,4- and 1,5-bisphosphates. The 1,4-bisphosphate was crystallised, phosphitylated and oxidized to either the fully protected trisphosphate or the fully protected 1,4-bisphosphate-5-phosphorothioate. Reductive deprotection with sodium in liquid ammonia gave either Ins(1,4,5)P$_3$ or Ins(1,4,5)P$_3$-5S respectively. 1-(2,2,2-Trichloroethyl) phospho-2,3,6-tri-*O*-benzyl-*myo*-inositol was used to prepare *myo*-inositol-1-phosphate-4,5-bisphosphorothioate in a similar fashion (N. J. Noble & B. V. L. Potter, unpublished).

The nucleophilic character of the sulfur atom of a phosphorothioate group makes this atom a suitable point of attachment for environmentally-sensitive reporter groups such as fluorescent, spin, affinity labels etc. This has been especially exploited in the oligonucleotide field. The first example of an inositol trisphosphate analogue modified by a reporter group involved the attachment of a fluorescent nitrobenzoxadiazole (NBD) moiety to the 1-phosphorothioate of *myo*-inositol 1-phosphate-4,5-bisphosphate [Figure 4, (4)] via the sulfur atom (10). The resulting NBD-Ins(1,4,5)P$_3$ analogue (6) was highly potent at releasing intracellular Ca^{2+} [6-fold weaker than Ins(1,4,5)P$_3$] and bound to the Ins(1,4,5)P$_3$ cerebellar receptor with high affinity.

Syntheses of other inositol phosphorothioates have been reported. Inositol 1-phosphorothioate has been prepared by thiophosphorylation of 1,2,4,5,6-penta-*O*-acetyl-*myo*-inositol and deblocking (15). *Endo-* and *exo-*diastereoisomers of the 5- membered (1,2-cyclic) phosphorothioates have been prepared by thiophosphorylation of 1,4,5,6-tetra-*O*-acetyl-*myo*-inositol (16) or by simultaneous mono-phosphitylation of 1,4,5,6-tetra-*O*-benzyl-*myo*-inositol at either the 1- or 2-positions, cyclisation and oxidation to the diastereoisomeric cyclic phosphorothioates, separation of diastereisomers and deblocking (17).

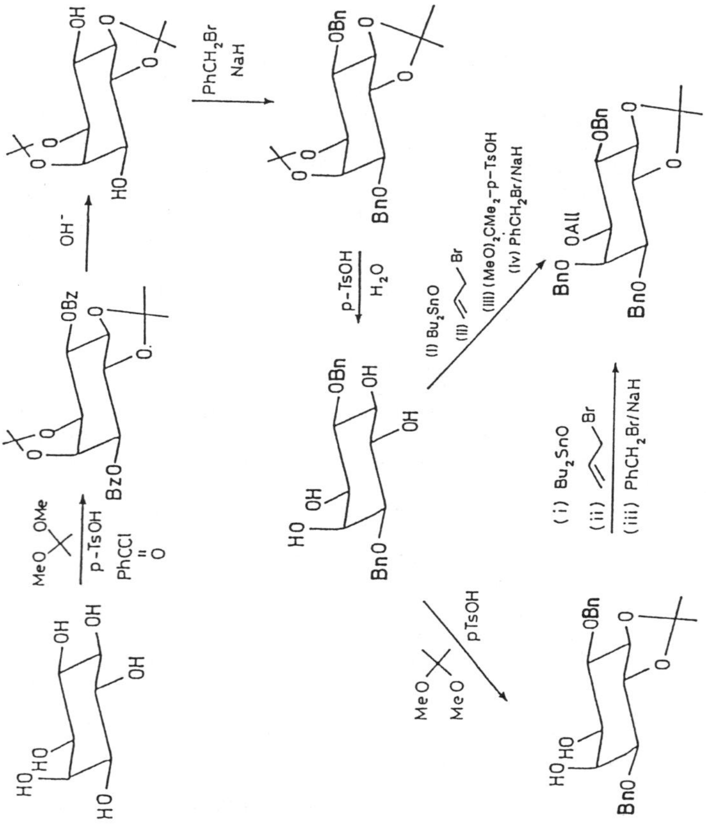
p-TsOH
PhCCl
OH⁻
PhCH2Br
NaH
p-TsOH
H2O
(i) Bu2SnO
(iii) (MeO)2CMe2-p-TsOH
(iv) PhCH2Br/NaH
pTsOH
(i) Bu2SnO
(iii) PhCH2Br/NaH

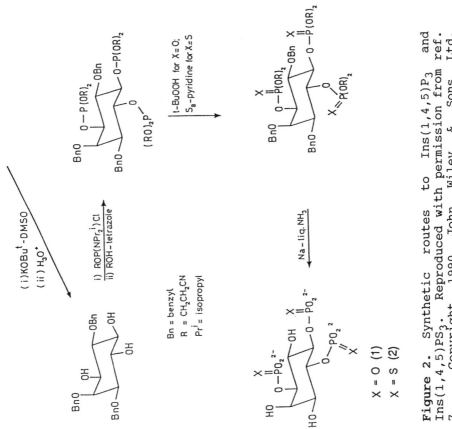

Figure 2. Synthetic routes to Ins(1,4,5)P$_3$ and Ins(1,4,5)PS$_3$. Reproduced with permission from ref. 7. Copyright 1990 John Wiley & Sons Ltd.

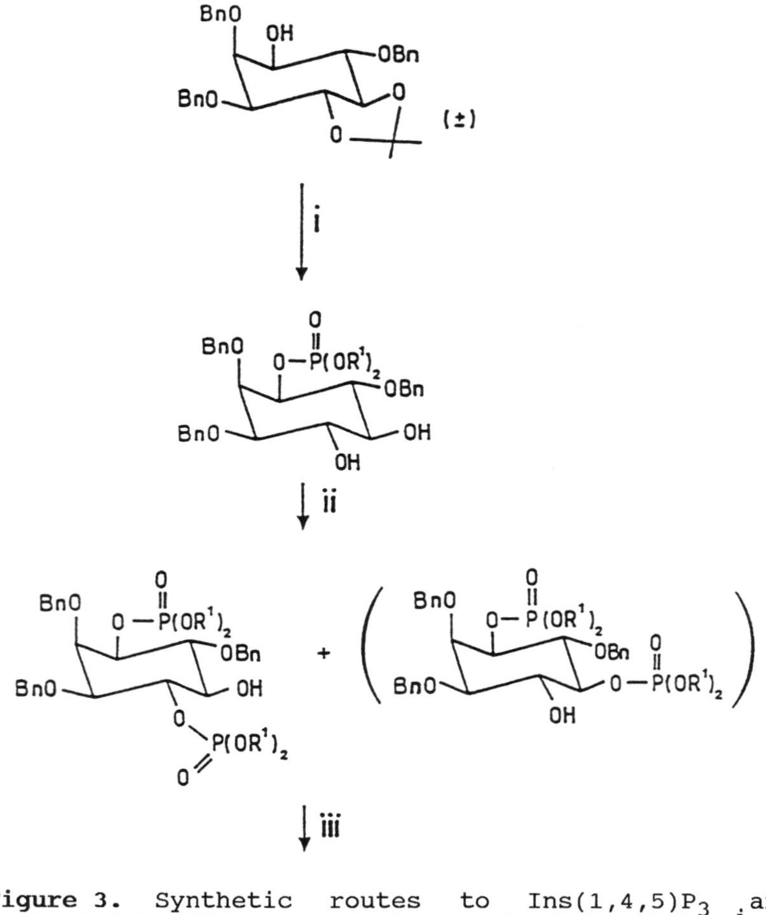

Figure 3. Synthetic routes to Ins(1,4,5)P$_3$ and Ins(1,4,5)P$_3$-5S. Reagents: i (a) ClP(NPr$^i_2$)-OCH$_2$CH$_2$CN, (b) NCCH$_2$CH$_2$OH-tetrazole, (c) for X=O, Bu^{t}OOH; X=S, sulphur in C$_5$H$_5$N; iv Na in liq. NH$_3$. Reproduced with permission from ref. 9. Copyright 1988 Royal Society of Chemistry, London.

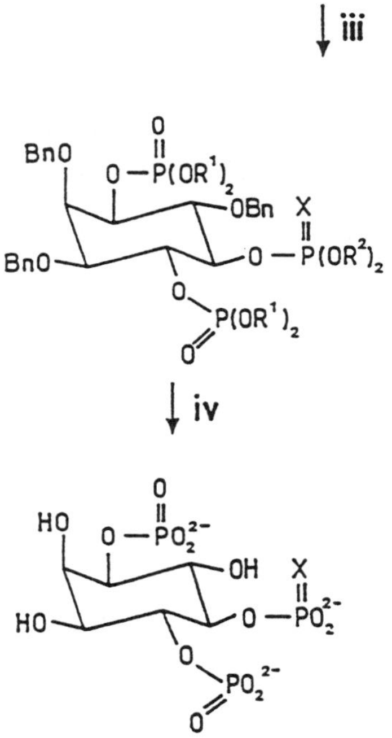

Figure 3. Continued.

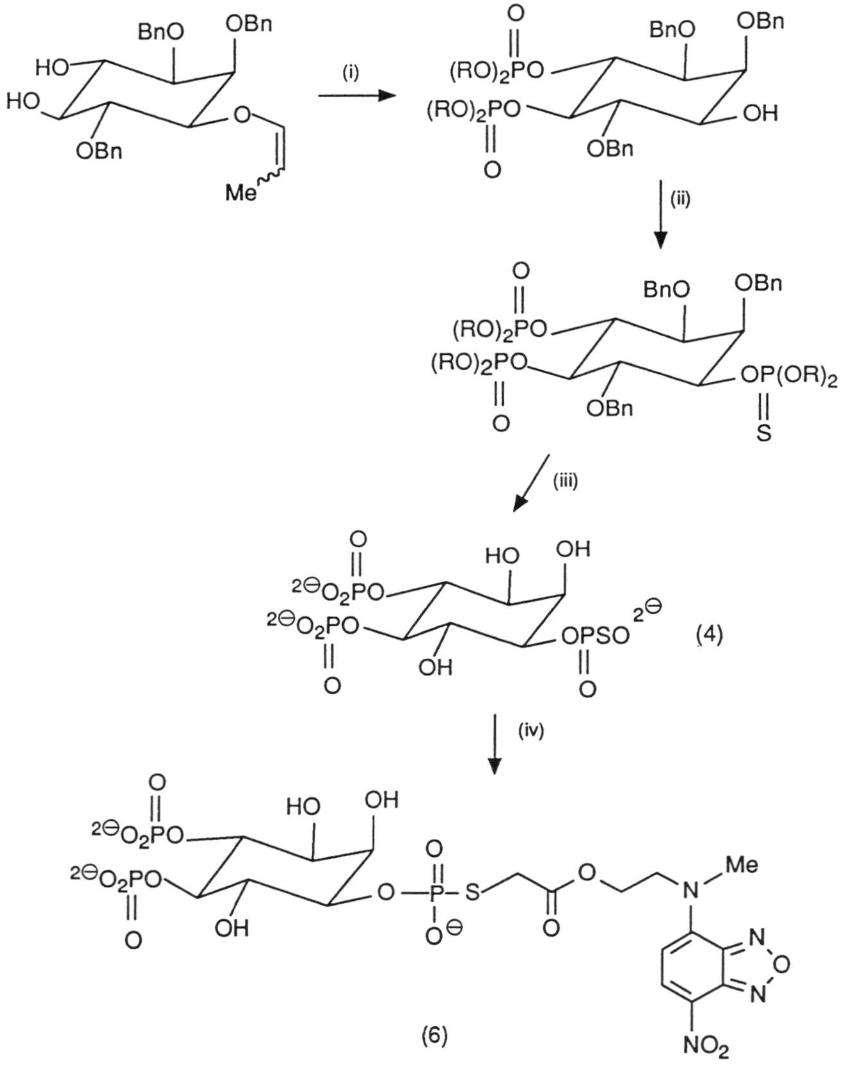

R = -CH₂CH₂CN

Bn=benzyl

Figure 4. Synthetic route to Ins(1,4,5)P₃-1S and NBD-analogue. Reagents: i (a) (RO)₂PNPriᵢ₂-tetrazole, (b) ButOOH, (c) HgO–HgCl₂; ii (a) (RO)₂PNPriᵢ₂-tetrazole, (b) sulphur in C₅H₅N; iii Na in liq. NH₃; iv IANBD-EtOH. All compounds are racemic. Reproduced with permission from ref. 10. Copyright 1990 Royal Society of Chemistry, London.

Biology and applications of inositol phosphates

Several studies on biological applications of $Ins(1,4,5)PS_3$ and $Ins(1,4,5)P_3$-5S have now been reported. Thus, inositol 1,4,5-trisphosphorothioate (2), binds to $Ins(1,4,5)P_3$ receptor sites in brain (12,18) and liver (19) and is only slightly less potent than $Ins(1,4,5)P_3$. Moreover, it is a full and potent agonist for the release of intracellular Ca^{2+} in a variety of systems, such as in *Xenopus* oocytes (20,21), permeabilised Swiss 3T3 cells (20,22), GH_3 cells (22), hepatocytes (23), pancreatic (24,25) and parotid (26) acinar cells, skeletal muscle traids (27), mouse lacrimal cells (28) and SH-SY5Y neuroblastoma cells (2), being only some 3-4 fold less potent than $Ins(1,4,5)P_3$. As expected from the properties of phosphorothioates, it is resistant to 5-phosphatase-catalyzed dephosphorylation (12,23) and can therefore produce a sustained calcium transient in cells (2,23). $Ins(1,4,5)PS_3$ is the most potent competitive inhibitor of 5-phosphatase yet reported (29), but it is not bound by the 3-kinase and does not compete with $Ins(1,4,5)P_3$ for this enzyme (23,30). Thus, $Ins(1,4,5)PS_3$ is recognised with high affinity by $Ins(1,4,5)PS_3$ receptors. It is a full agonist with respect to Ca^{2+} release and yet is resistant to all known routes of $Ins(1,4,5)P_3$ metabolism.

The possible synergy between $Ins(1,4,5)P_3$ and $Ins(1,3,4,5)P_4$ is promoting Ca^{2+} entry at the plasma membrane has been the subject of considerable debate. In singly internally perfused mouse lacrimal acinar cells, using the patch clamp technique for whole cell current recording, monitoring a Ca^{2+}-activated K^+ current, $Ins(1,4,5)PS_3$ alone gives rise to a single transient response, typical of $Ins(1,4,5)P_3$, and independent of external Ca^{2+}. Together with $Ins(1,3,4,5)P_4$, however , it evokes a sustained response dependent upon external Ca^{2+}, suggesting that the transient response of $Ins(1,4,5)P_3$ is not a consequence of rapid metabolism, and that $Ins(1,3,4,5)P_4$ is not acting by protecting $Ins(1,4,5)P_3$ against dephosphorylation by the common 5-phosphatase (28).

Agonist-stimulated cells often give rise to oscillating internal Ca^{2+} levels, rather than a smooth rise, and the role and mechanism of generation of such oscillations have been the subject of significant interest and controversy. In singly internally perfused mouse pancreatic acinar cells $Ins(1,4,5)PS_3$, applied through a patch pipette, evokes repetitive pulses of Ca^{2+}-activated Cl^- current, which are similar in amplitude and frequency to the response of such cells to acetylcholine, acting through muscarinic receptors. Thus, pulsatile Ca^{2+} release is possible even at a constant level of this analogue, and presumably, therefore, of Ins(1,4,5)P$_3$. This has been used to argue against a receptor-controlled oscillator in the generation of Ca^{2+} oscillations, as well as any role for the periodic phosphorylation or degradation of Ins(1,4,5)P$_3$ (25).

The polyphosphoinositide pathway is thought to mediate the ability of light to release Ca^{2+} from ER within invertebrate microvillar photoreceptors, via the formation of Ins(1,4,5)P$_3$. Ins(1,4,5)PS$_3$ has been used to investigate mechanisms that terminate the mobilisation of Ca^{2+} in ventral photoreceptors of the horseshoe crab Limulus. It can generate sustained repetitive oscillations of Ca^{2+}-dependent membrane potential in the Limulus photoreceptor, where the action of Ins(1,4,5)P$_3$ is normally rapidly terminated by metabolism (31). Ins(1,4,5)PS$_3$ is also capable of generating oscillations of membrane potential (20) and Ca^{2+}-dependent Cl$^-$ current (21) in Xenopus oocytes. In the oocyte, however, such oscillations are not sustained, indicating that factors other than metabolism are important in terminating the response. Oscillations of membrane potential caused by Ins(1,4,5)PS$_3$ are different from those generated by Ins(1,4,5)P$_3$ and resemble more those from Ins(2,4,5)P$_3$. The exact reason for this is not yet clear, but different mechanisms for setting up Ca^{2+} oscillations may be possible (32). Oscillations in Ca^{2+}-dependent Cl$^-$ current, induced by Ins(1,4,5)PS$_3$, however, appear to resemble those induced by Ins(1,3,4,5)P$_4$ rather than Ins(1,4,5)P$_3$ (33).

Ins(1,4,5)PS$_3$ has been used in rat pancreatic acinar cells to help distinguish functionally between Ins(1,4,5)P$_3$-sensitive and -insensitive non-mitochondrial MgATP-dependent Ca^{2+} pools (24). Ins(1,4,5)PS$_3$ was used to keep the Ins(1,4,5)P$_3$-sensitive Ca^{2+} pool empty and Ca^{2+} reuptake occurred into the Ins(1,4,5)P$_3$ insensitive pool. However, in experiments on Ca^{2+} mobilisation in permeabilised rat parotid acinar cells, evidence has been obtained for Ca^{2+} reuptake into an Ins(1,4,5)P$_3$- and thapsigargin-sensitive Ca^{2+} store in the presence of Ins(1,4,5)PS$_3$ (26).

By virtue of their properties as potent 5-phosphatase inhibitors, Ins(1,4,5)P$_3$ and the related monophosphorothioate analogue, inositol 1,4-bisphosphate-5-phosphorothioate Ins(1,4,5)P$_3$-5S (9), have been employed to inhibit Ins(1,4,5)P$_3$ breakdown in electrically-permeabilized SH-SY5Y human neuroblastoma cells (34). Inhibition of 5-phosphatase-mediated metabolism of exogenously added 5[^{32}P]-Ins(1,4,5)P$_3$ was ca. 10 times greater than that of cell membrane-derived [^{3}H]-Ins(1,4,5)P$_3$, indicating the possibility of Ins(1,4,5)P$_3$ compartmentation, i.e. that homogenous mixing of exogenously-added and endogenously-generated Ins(1,4,5)P$_3$ does not occur.

Another application of $Ins(1,4,5)PS_3$ has been in the investigation of 'quantal' release of intracellular Ca^{2+} by $Ins(1,4,5)P_3$ in permeabilized hepatocytes (*35*), where the size of the $Ins(1,4,5)P_3$-sensitive Ca^{2+} pool is apparently dependent upon the concentration of $Ins(1,4,5)P_3$. The failure of sub-maximal concentrations of $Ins(1,4,5)P_3$ or $Ins(1,4,5)PS_3$ to empty the Ca^{2+} store completely was not due to deactivation of the stimulus or receptor desensitisation. The metabolic stability of $Ins(1,4,5)P_3$ allowed Ca^{2+} efflux experiments to be performed at a high cell density where degradation of $Ins(1,4,5)P_3$ would normally have posed significant problems.

This stability of $Ins(1,4,5)PS_3$ was also crucial for the observation that an analogue of $Ins(1,4,5)PS_3$ can activate a novel voltage-dependent K^+ conductance in rat CA1 hippocampal pyramidal cells (*36*). $Ins(1,4,5)PS_3$ inhibited action potential firing when injected into these cells. $Ins(1,4,5)P_3$ itself did not elicit this conductance, presumably because of its rapid breakdown in these cells. Thus, use of $Ins(1,4,5)PS_3$ may uncover activities of $Ins(1,4,5)P_3$ which may not be experimentally observable using the natural messenger because of rapid metabolism or slow diffusion of exogenously-added messenger.

$Ins(1,4,5)PS_3$ released up to 20% of an actively-loaded Ca^{2+} pool in triads from rabbit skeletal muscle, although activation of ryanodine receptor Ca^{2+} channels was zero or minimal (*27*), raising the possiblility that the Ca^{2+} mobilising activity may be mediated by other channels or mechanisms. The kinetics of Ca^{2+} release by $Ins(1,4,5)PS_3$ in the sarcoplasmic reticulum of skeletal muscle was surprisingly fast (20-100 ms, depending upon agonist concentration), indicating that $Ins(1,4,5)P_3$ receptors of skeletal muscle are kinetically competent to signal the rapid elevation of cytosolic Ca^{2+} that precedes muscle contraction (*37*).

Although polyphosphoinositides turn over in human red blood cells, a role for $Ins(1,4,5)P_3$ has yet to be established. In permeabilised human red blood cells $Ins(1,4,5)P_3$ evokes sustained release of Ca^{2+} and irreversible shape changes and disorganisation of the spectrin network, as measured by immunofluorescence, whereas $Ins(1,4,5)P_3$ evokes reversible shape changes (*38*). The polyphosphoinositide signalling pathway evidently plays an important role in the shape maintenance of red blood cells.

[^{35}S]-Labelled myo-inositol phosphorothioates

The wide range of applications of inositol phiosphorothioates detailed above has been extended by the commercial availability of [^{35}-S]-labelled $Ins(1,4,5)PS_3$, which has been shown to be a valuable metabolically stable radioligand with high affinity for the $Ins(1,4,5)P_3$ receptor (39).

[35S]-Labelled $D-Ins(1,4,5)PS_3$ was prepared by an adaptation of the chemical synthesis of unlabelled material (8). Experiments to characterize the interaction of this ligand with the cerebellar $Ins(1,4,5)P_3$ receptor have shown that $D-[^{35}S]-Ins(1,4,5)PS_3$ binds with high affinity (K_d, 31 ± 2.8 nM) to cerebellar membranes, revealing a high density of $Ins(1,4,5)P_3$ receptors to be present (B_{max} 16.9 ± 0.3 pmol/mg protein). Identical results are obtainable by saturation analysis of the isotope dilution data obtained by competition with unlabelled $Ins(1,4,5)PS_3$. These results agree with previous estimates of $Ins(1,4,5)P_3$ receptor density in this tissue (18) and suggest that $[^{35}S]-Ins(1,4,5)PS_3$ has only a slightly lower affinity for this binding site, consistent with a study on the unlabelled ligand (12).

Further studies suggest that the association rate of $[^{35}S]-Ins(1,4,5)PS_3$ with the cerebellar receptor site is rapid, with equilibrium binding being attained within 10 min. at $4°C$ and that the stereo- and positional displacement of specifically bound $[^{35}S]-Ins(1,4,5)PS_3$ by $D-Ins(1,4,5)P_3$, $L-Ins(1,4,5)P_3$ and $D-Ins(2,4,5)P_3$ have the same rank order compared to previous studies using $D-[^3H]-Ins(1,4,5)P_3$ as the radioligand (18). The real potential value of this radioligand lies with its metabolic stability, allowing evaluation of the properties of the $Ins(1,4,5)P_3$ receptor at $37°C$ and with a physiological ionic environment.

The preparation of radiolabelled phosphorothioate analogues of $Ins(1,4,5)P_3$ has also been achieved by phosphorylation of the polyphosphoinositide lipids PtdIns and $PtdIns(4)P$ using human erythrocyte ghost kinases and $ATP\gamma S$. $[^{35}S]$-Radiolabelled material can be synthesized by employing $[^{35}S]-ATP\gamma S$, and incubation of $[^{35}S]-ATP\gamma S$ with erythrocyte ghosts produced $[^{35}S]$-labelled $PtdIns(4,5)P_2$ analogues, with $[^{35}S]$-label uniquely in the 5-position or in both the 4- and 5-positions [Figure 5]. These modified lipids were cleaved by the endogenous Ca^{2+}-activated phosphoinositidase C to give a mixture of inositol 1,4-bisphosphate-5[35S]-phosphorothioate (7) and inositol 1-phosphate-4,5[35S]-bisphosphorothioate (8), which was demonstrated to be resistant to 5-phosphatase. (40). Chemical synthesis of $[^{35}S]$-inositol phosphorothioates as described above, however, offers a more direct and reproducible strategy.

In summary, inositol phosphorothioates are proving to be valuable pharmacological and biochemical tools. Clearly, the commercial availability of both $D-Ins(1,4,5)PS_3$ and $D-[^{35}S]-Ins(1,4,5)PS_3$ will now provide even greater opportunities for the biological exploitation of these novel analogues.

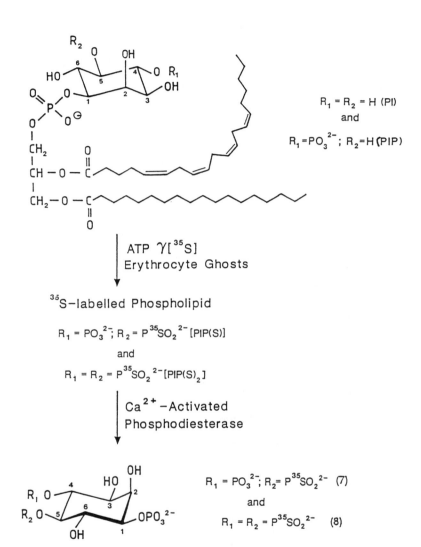

Figure 5. Preparation of [^{35}S]-inositol phosphorothioates from PtdIns and PtdIns(4)P.   After Folk *et al.* (*40*).   Reproduced with permission from ref. 7. Copyright 1990 John Wiley & Sons Ltd.

Acknowledgments

Research on *myo*-inositol phosphorothioates was supported by the Science & Engineering Research Council (U.K.). [^{35}S]-labelled *myo*-inositol phosphorothioates were synthesized by Dupont NEN (Boston, U.S.A.). Biological evaluations were carried out in collaboration with S. R. Nahorski. We thank Susan Alston for manuscript preparation. B. V. L. Potter is a Lister Institute Research Professor.

Literature cited

1.   Berridge, M. J. *Annu. Rev. Biochem.* **1987**, *56*, 159-193.
2.   Nahorski, S. R.; Potter, B. V. L. *Trends Pharmacol. Sci.* **1989**, *10*, 139-144.
3.   Chilvers, E. R.; Kennedy, E. D.; Potter, B. V. L. *Drug News & Perspectives* **1989**, *2*, 342-346.
4.   Ghosh, T. K.; Eis, P. S.; Mullaney, J. M.; Ebert, C. L.; Gill, D. L. *J. Biol. Chem.* **1988**, *263*, 11075-11079.
5.   Billington, D. C. *Chem. Soc. Rev.* **1989**, *18*, 83-122.
6.   Potter, B. V. L. *Nat. Prod. Reports* **1990**, *7*, 1-24.
7.   Potter, B. V L. In *Transmembrane Signalling, Intracellular Messengers & Implications for Drug Development*, Nahorski, S. R., ed.; Wiley, Chichester, UK., 1990, pp207-239.
8.   Cooke, A. M.; Noble, N. J.; Gigg, R.; Payne, S.; Potter, B. V. L. *J. Chem. Soc. Chem. Commun.* **1987**, 1525-1526.
9.   Cooke, A. M.; Noble, N. J.; Gigg, R.; Payne, S.; Potter, B. V. L. *J. Chem. Soc. Chem. Commun.* **1988**, 269-271.
10.   Lampe, D.; Potter, B. V. L. *J. Chem. Soc. Chem. Commun.* **1990**, 1500-1501.
11.   Hamblin, M. R.; Flora, J. S.; Potter, B. V. L.; *Biochem. J.* **1987**, *246*, 771-774.
12.   Willcocks, A. L.; Potter, B. V. L.; Cooke, A. M.; Nahorski, S. R. *Eur. J. Pharmacol.* **1988**, *155*, 181-183.
13.   Hamblin, M. R.; Gigg, R.; Potter, B. V. L. *J. Chem. Soc. Chem. Commun.* **1987**, 626-627.
14.   Gigg, J.; Gigg, R.; Payne, S.; Conant, R. *J. Chem. Soc. Perkin Trans. I* **1987**, 423-429.
15.   Metschies, T.; Schulz, C.; Jastorff, B. *Tet. Lett.* **1988**, *29*, 3921-3922.
16.   Schulz, C.; Metschies, T.; Jastorff, B. *Tet. Lett.* **1988**, *29*, 3919-3920.
17.   Lin. G.; Tsai, M. -D. *J. Amer. Chem. Soc.* **1989**, *111*, 3099-3101.
18.   Willcocks, A. L.; Cooke, A. M.; Potter, B. V. L.; Nahorski, S. R. *Biochem. Biophys. Res. Commun.* **1987**, *146*, 1071-1078.

19. Nunn, D. L.; Potter, B. V. L.; Taylor, C. W. *Biochem. J.* **1990**, *265*, 393-398.
20. Taylor, C. W.; Berridge, M. J.; Brown, K. D.; Cooke, A. M.; Potter, B. V. L. *Biochem. Biophys. Res. Commun.* **1988**, *150*, 626-632.
21. DeLisle, S.; Krause, K. -H.; Denning, G.; Potter, B. V. L.; Welsh, M. J. *J. Biol. Chem.* **1990**, *265*, 11726-11730.
22. Strupish, J.; Cooke, A. M.; Potter, B. V. L.; Gigg, R.; Nahorski, S. R. *Biochem. J.* **1988**, *253*, 901-905.
23. Taylor, C. W.; Berridge, M. J..; Cooke, A. M.; Potter, B. V. L. *Biochem. J.* **1989**, *259*, 645-650.
24. Thevenod, F.; Dehlinger-Kremer, M.; Kemmer, T. P.; Christian, A. -L.; Potter, B. V. L.; Schulz, I. *J. Membr. Biol.* **1989**, *109*, 173-186.
25. Wakui, M.; Potter, B. V. L.; Petersen, O. H. *Nature* **1989**, *339*, 317-320.
26. Mennitti, F. S.; Takemura, H..; Thastrup, O.; Potter, B. V. L.; Putney, J. W., Jnr. *J. Biol. Chem.* **1991** submitted.
27. Valdivia, C.; Valdivia, H. H.; Potter, B. V. L.; Coronado, R. *Biophys. J.* **1990**, *57*, 1233-1243.
28. Changya, L.; Gallacher, D. V.; Irvine, R. F.; Potter, B. V. L.; Petersen, O. H. *J Membr. Biol.* **1989**, *109*, 85-93.
29. Cooke, A. M.; Nahorski, S. R.; Potter, B. V. L. *FEBS Lett.* **1989**, *242*, 373-377.
30. Safrany, S. T.; Wojcikiewicz, R. J. H.; Strupish, J.; McBain, J.; Cooke, A. M.; Potter, B. V. L.; Nahorski, S. R. *Brit. J. Pharmacol. Proc. Suppl.* **1990**, *99*, 88P.
31. Payne, R. F.; Potter, B. V. L. *J. Gen. Physiol.* **1991**, in press.
32. Berridge, M. J.; Potter, B. V. L. *Cell Regulation* **1990**, *1*, 675-681.
33. Ferguson, J.; Potter, B. V. L.; Nuccitelli, R. *Biochem. Biophys. Res. Commun.* **1990**, *172*, 229-236.
34. Wojcikiewicz, R. J. H.; Cooke, A. M.; Potter, B. V. L.; Nahorski, S. R. *Eur. J. Biochem.* **1990**, *192*, 459-467.
35. Taylor, C. W.; Potter, B. V. L.; *Biochem. J.* **1990**, *266*, 189-194.
36. McCarren, M.; Potter, B. V. L.; Miller, R. J. *Neuron* **1989**, *3*, 461-471.
37. Valdivia, C.; Vaughan, D.; Potter, B. V. L.; Coronado, R. *Science* **1991** submitted.
38. Strunecka, A.; Kmonickova, E.; El Desouki, N.; Krpejsova, L.; Palacek. J.; Potter, B. V. L. *Receptor* **1991**, in press.
39. Potter, B. V. L.; Challiss, R. A. J.; Nahorski, S. R. *Dupont Biotech Updates* **1990**, *5*, 85-90.
40. Folk, P.; Kmonickova, E.; Krpejsova, L.; Strunecka, A. *J. Labelled Comp. Radiopharm.* **1988**, *25*, 793-803.

RECEIVED February 11, 1991

Chapter 15

Synthesis and Biological Evaluation of Inositol Derivatives as Inhibitors of Phospholipase C

Franz Kaufmann, D. James R. Massy[1], Wolfgang Pirson, and Pierre Wyss[2]

Pharmaceutical Research Department, F. Hoffmann–La Roche Limited, CH–4002 Basel, Switzerland

The synthesis and biological evaluation of analogues of phosphatidylinositol (PI) and phosphatidylinositol 4,5-bisphosphate (PIP$_2$) is described. Inositol derivatives with and without homologation at C(1) and with and without ionic groups (phosphate or sulfate) at C(4) and C(5) were prepared. In all these compounds, palmitate ester groups were introduced in place of the diacylglyceryl group of PI or PIP$_2$. All compounds with ionic groups (sulfates or phosphates) were found to inhibit phospholipase C (PLC) *in vitro* significantly. In contrast, compounds without ionic groups showed only a slight degree of inhibition. Some of the compounds with ionic groups have also been tested for PLC inhibition in intact cells and were found to be inactive in this assay.

Introduction.- A number of studies have documented the pivotal role of *myo*-inositol 1,4,5-trisphosphate (IP$_3$) and diacylglycerol (DG) in the intracellular signalling pathway [1-4]. Thus, it has been established that extracellular signals can trigger the hydrolysis of phosphatidylinositol 4,5-bisphosphate (PIP$_2$) to IP$_3$ and DG, a reaction which is catalysed by phospholipase C (PLC) in the cell membrane. It seems likely that this initial event is associated with both inflammatory responses *via* the arachidonic-acid cascade [5] and the regulation of cell proliferation [6]. As part of a programme aimed at obtaining substances controlling the functions of the second messengers mentioned above, we have synthesised a number of compounds, analogues of both phosphatidylinositol (PI) and PIP$_2$, as potential inhibitors of PLC. All compounds described are either *meso* or racemic mixtures. In the latter case the structures portrayed here correspond, for convenience, to one enantiomer, but the DL-form is implied. Numbering and nomenclature are in accordance with IUPAC conventions [7], except for the abbreviations PI, PIP$_2$, IP$_3$, and DG which are based on the 1984 Chilton Conference system [8].

[1]Current address: School of Chemical Sciences, University of East Anglia, Norwich NR4 7TJ, England
[2]Corresponding author

0097–6156/91/0463–0202$06.00/0

We report here the synthesis and biological evaluation of PI and PIP$_2$ analogues in which a CH$_2$ group isosterically replaces the oxygen attached to C-1 of the inositol ring, and the lipid moiety is replaced by a palmitoyloxy group attached to this CH$_2$ group. The function of PLC being to cleave a lipid/phosphate bond, we anticipated that such homologated inositol carboxylates would be effective inhibitors. To ascertain the importance of the PIP$_2$ phosphate groups at C(4) and C(5), we prepared and evaluated C(1) homologated *myo*-inositol palmitates both with and without ionic groups (phosphates or sulfates) at C(4) and C(5), (**4, 5** and **6**). The corresponding *chiro*-inositol compounds were also synthesised and evaluated (**1, 2** and **3**) as well as two nonionic deoxy *myo*-inositol compounds (**7** and **8**). Finally, *myo*-inositol palmitates with ionic groups at C(4) and C(5) but without homologation at C(1), were prepared for comparison (**9** and **10**).

Biological   evaluation.- *1. Inhibition of phospholipase C (PLC)* in vitro. Phospholipase C was prepared from human platelets as crude cytosolic enzyme obtained after sonication and centrifugation (100000 x g supernatant) of platelets isolated from fresh human blood. This enzyme preparation was stored in aliquots at -75° C. Phospholipase C activity was determined by measuring the hydrolysis of phos-

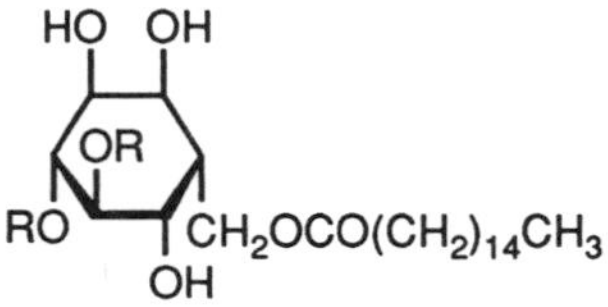

HO   OH

OR

RO   CH₂OCO(CH₂)₁₄CH₃
     OH

1   R = H

2   R = SO₃Na

3   R = P(O)(OH)₂ · 1.6 Me₃N

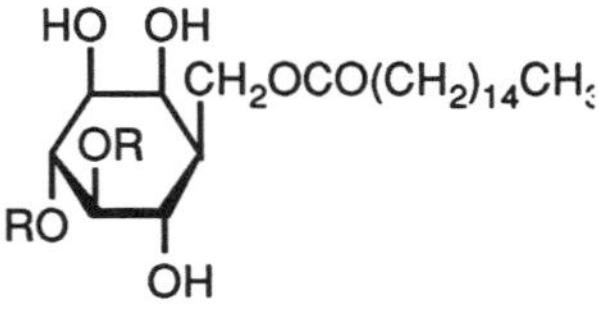

HO   OH
        CH₂OCO(CH₂)₁₄CH₃
OR

RO
        OH

4   R = H

5   R = SO₃Na

6   R = P(O)(OH)₂ · 1.6 Me₃N

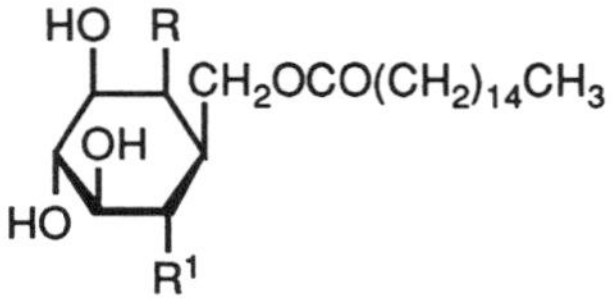

HO   R
        CH₂OCO(CH₂)₁₄CH₃
OH

HO
     R¹

7   R = H,   R¹ = OH

8   R = OH, R¹ = H

HO   OH
        OCO(CH₂)₁₄CH₃
OR

RO
        OH

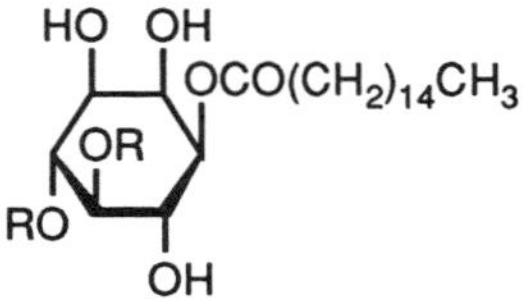

9   R = SO₃Na

1 0   R = P(O)(OH)₂ · 2.5 Me₃N

phatidyl-[³H]inositol (PI) as a substrate at pH 5.5 without any detergent. Water-soluble ³H-inositolphosphate was determined as product in the water phase after extraction of ³H-PI with chloroform. The assay was carried out in 0.1 ml for 10 min at 37°C in the presence of 100 mM Tris-acetate, pH 5.5, 0.5 mM CaCl₂ and 250 µM sonicated ³H-PI. The substrate suspension was prepared by mixing ³H-phosphatidyl-inositol (³H-PI from Amersham) and unlabelled PI (Sigma), removing the solvent chloroform under N₂, and sonicating the residue in 200 mM Tris-acetate buffer, pH 5.5, containing 1 mM CaCl₂ for 20 min. For the assay, 10 ml DMSO (with or without test compound) was mixed with 40 µL of enzyme solution (containing about 10 µg protein). After preincubation for 15 min the reaction was started with 50 µL substrate suspension. The assay was run for 10 min. Thereafter the reaction was stopped by addition of 1.5 mL of an ice-cold mixture of chloroform/butanol/conc. HCl (10:10:0.6) and 0.45 mL of 1 N HCl. After vigorous mixing the phases were separated by centrifugation for 10 min (1000 rpm). An aliquot of 0.2 mL from the upper water phase was counted for radioactivity. Under the assay conditions 10-20% of the PI was hydrolyzed. In this range the assay was linear with time and enzyme. For each concentration of the test compound the percent inhibition was calculated relative to controls with vehicle only. The concentration for 50% inhibition (IC₅₀) was determined graphically.

The synthetic compounds described above were evaluated according to the procedure just described. The results are summarized in the following Table. The compounds without ionic groups showed only a slight degree of inhibition. In contrast, all compounds with ionic groups (sulfates or phosphates) were found to

Table. Evaluation of Inositol Derivatives

	Compound	PLC Inhibition	
		%, 500 μM*	IC$_{50}$ (μM)
1	(nonionic)	39	
4	"	27	
7	"	26	
8	"	17	
2	(ionic)		30
3	"		14.5
5	"		22
6	"		14
9	"		18
10	"		230

* Inhibition too feeble for determination of IC$_{50}$.

inhibit the enzyme significantly. The results, however, do not seem to show any other significant correlation between structure and activity.

2. *Formation of inositol phosphates in GH3 cells.* Some of the compounds have been tested for PLC inhibition in intact cells by measuring their effect on the formation of inositol phosphates in GH3 cells on TRH (Thyrotropin releasing hormone) induction. ^{3}H-Inositol labelled cells were incubated for 30 min with a solution of the test compound in dimethylsulfoxide (DMSO), followed by stimulation with TRH (25 nM) for 30 min at 37°C. The maximal DMSO concentration in the incubation medium was 0.1%. The ^{3}H-inositol phosphates liberated in the cell were extracted and determined by anion exchange chromatography as described elsewhere [9].

The compounds tested in this assay (**3**, **5**, **6** and **10**) did not inhibit the formation of inositol phosphates up to a concentration of 10 mM. We suspect that the lack of activity in this system might be due to a poor penetration of the charged compounds into the cells. In order to overcome this potential problem, we are now concentrating our efforts in the synthesis of prodrugs.

Chemistry. Full details on the synthetic work discussed here have recently been published [10].

1. *Homologation at C(1) without Ionic Groups in the Molecule.* The initial target of this work was the homologation of inositol at C(1) without the introduction of ionic groups, *i.e.* compounds **1** and **4** . These compounds which are analogues of PI were prepared by the route shown in *Scheme 1*. Thus, *myo*-inositol was reacted with excess cyclohexanone in DMF/toluene, catalysed by TsOH, to give the monocyclohexylidene compound **11**. Benzylation of **11** yielded the tetrabenzyl compound **12**, which was hydrolysed to give *cis*-diol **13**. We adopted the following route to convert this to the desired 1-alcohol **16**: stannylidene-activated allylation [11] (**13**---->**14**), benzylation (**14**---->**15**), and deallylation of **16** to ketone **17** proceeded satisfactorily by the use of pyridinium chlorochromate. Homologation of **17** by means of a *Wittig* reaction, followed by hydroboration of the methylidene compound **18** afforded the epimeric alcohol mixture **19**. This mixture was readily esterified with palmitoyl chloride, and the resulting esters **20** and **21** could be separated cleanly by MPLC. The ratio of *chiro*-ester **20** to *myo*-ester **21** was *ca.* 2.5:1. The corresponding deprotected compounds **1** and **4** were obtained as pure, crystalline compounds.

Ketone **17** was reacted with (methoxymethyl)triphenylphosphonium chloride to yield enol ethers **22a** and **22b** (*Scheme 2*). Conversion of these enol ethers to the

Scheme 1

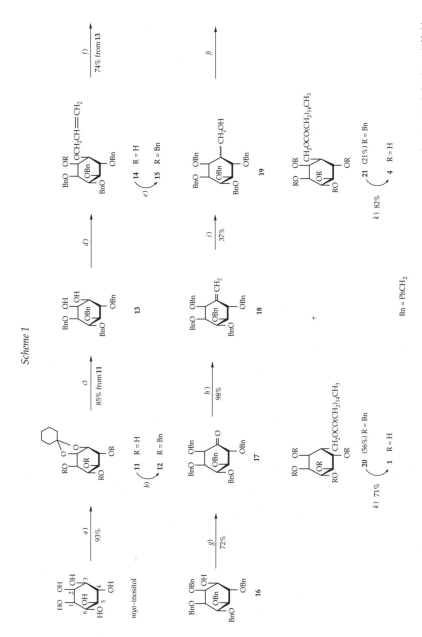

Bn = PhCH₂

a) 1) Cyclohexanone, TsOH, DMF/toluene, 144° 9 h; 2) EtOH, TsOH, 25° 1 h. *b*) PhCH₂Cl, KOH, 100-140° 1 h. *c*) 80% AcOH, 80-100° 1½ h. *d*) 1) Bu₂SnO, toluene, 110° 1 h; 2) CH₂=CHCH₂Br, NaH, DMF, 80° 4 h. *e*) PhCH₂Br, NaH, DMF, 40° 4 h. *f*) 5% Pd/C, TsOH, EtOH/ H₂O, reflux, 22 h. *g*) Pyridinium chlorochromate, CH₂Cl₂, 25° 48 h. *h*) CH₃(Ph)₃PBr, BuLi, THF, 1-25° 1 h. *i*) 1) BH₃·Me₂S, toluene, 60° 1 h; 2) 5M NaOH, 30% H₂O₂, 50° 1 h. *j*) CH₃(CH₂)₁₄COCl, pyridine, 4-(dimethylamino)pyridine, CH₂Cl₂, 25°. *k*) Pd/C, 1 atm H₂, EtOH, 25° 2-3 h.

Scheme 2

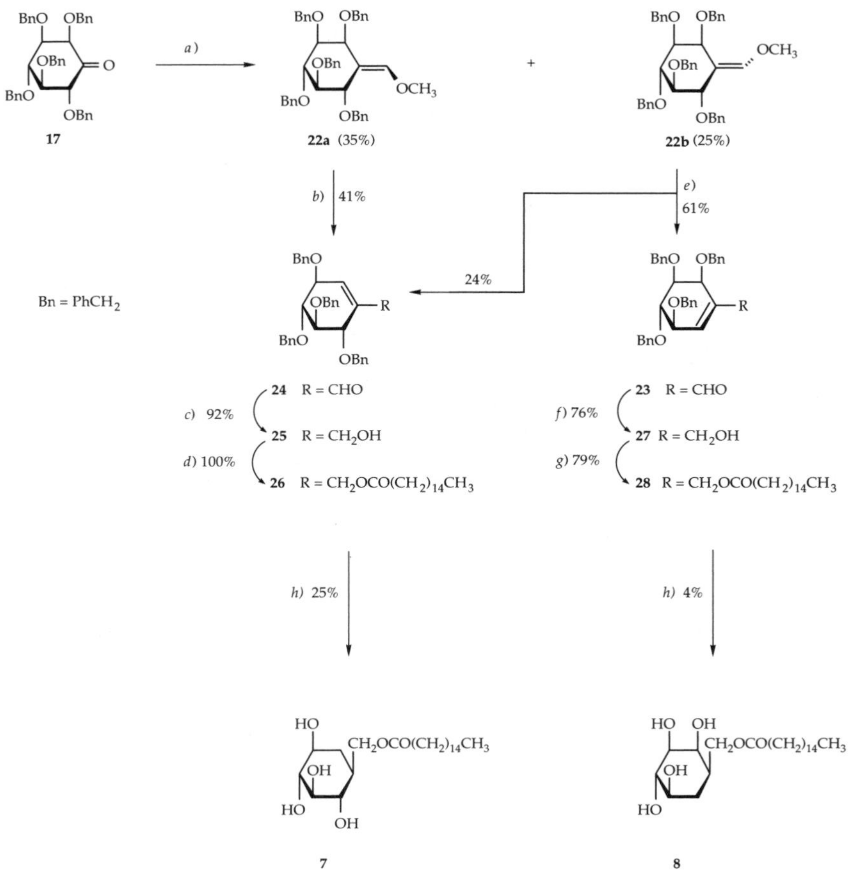

a) (CH₃OCH₂)P(Ph)₃Cl, BuLi, THF, 0-21°, ³/₄ h. b) CH₂Cl₂ (trace of adventitious HCl), 25°, 4 d. c) NaBH₄, EtOH, 25°, 25 min. d) CH₃(CH₂)₁₄COCl, pyridine, 4-(dimethylamino)pyridine, CH₂Cl₂, 25°, 2 h. e) BF₃·Et₂O, Et₂O, reflux, 1¹/₂ h. f) NaBH₄, EtOH, 25°, 10 min. g) CH₃(CH₂)₁₄COCl, Et₃N, 4-(dimethylamino)pyridine, CH₂Cl₂, 25°, 10 min. h) Pd/C, 1 atm H₂, EtOH, 25°, 2-4 h.

corresponding aldehydes was accompanied by the concomitant loss of a benzyloxy group and introduction of ring insaturation. In the case of **22a**, this transformation occurred spontaneously in CH_2Cl_2 solution at r.t. for a few days to give the unsaturated aldehyde **24**. The reaction appeared to be catalysed by traces of HCl present in the solvent since no appreciable decomposition was observed in EtOH. In contrast, enol ether **22b** did not decompose in CH_2Cl_2 solution, but on refluxing it in Et_2O with BF_3, the unsaturated aldehydes **23** and **24** were obtained in a *ca.* 2.5:1 ratio. Having a reasonable route to unsaturated aldehydes **23** and **24**, we converted them to target compounds of the deoxy type (**7** and **8**). Thus, reduction of aldehydes **23** and **24** using $NaBH_4$ gave alcohols **25** and **27**, which were then converted to esters **26** and **28**, respectively. Finally, these esters were converted to the target compounds by concomitant hydrogenolysis of the benzyl groups and reduction of the double bonds. The obtained deoxy compounds **7** and **8** are derivatives of compounds known as pseudo-sugars, *i.e.* sugars in which the ring O-atom is replaced by CH_2.

    2. Homologation at C(1) with Ionic Groups at C(4) and C(5). The route followed is shown in *Schemes 3* and *4* . The synthetic pathway requires temporary protection of the C(4) and C(5) OH groups of *myo*-inositol in two stages. In the first of these, protection was accomplished by forming the 1,2:4,5-di-O-cyclohexylidene compound **29**. The diketal **29** was then dibenzylated (**30**) after which the OH groups at C(4) and C(5) were re-exposed by selective removal of one cyclohexylidene group using an exchange reaction with ethane-1,2-diol. The OH groups at C(4) and C(5) in diol **31** were re-protected by 4-methoxybenzyl groups. The ketal group of **32** was then carefully hydrolysed. Selective benzylation of the axial OH-C(2) of **33** was accomplished by the allyl-ether technique described above (*Scheme 1*), except that here it was necessary to perform the isomerisation of the benzylated allyl ether **35** to the 1-propenyl ether **36** and hydrolytic removal of the propenyl-ether group in separate stages in order to maintain reaction conditions which would preserve the 4-methoxybenzyl groups. Oxidation of **37** to ketone **38**, followed by homologation and hydroboration gave the epimeric mixture **40**. The alcohol mixture **40** was separated after esterification and removal of the 4-methoxybenzyl groups yielding **42a** and **42b**, the *chiro-* and *myo*-type compounds.

    The diols **42a** and **42b** were converted to the corresponding sulfates **43** and **45** using SO_3/pyridine complex in DMF. Removal of the benzyl groups by hydrogenolysis yielded the target compounds **2** and **5** (*Scheme 4*). Phosphorylation of diol **42a** was achieved with bis(benzyloxy)(diisopropylamino)phosphine [12, 13] in the presence of $1H$-tetrazole, followed by oxidation of the phosphite ester to phosphate using 3-chloroperbenzoic acid [14] (*Scheme 4*). The heptabenzyl compound **44**, thus obtained, was then fully deprotected by hydrogenolysis and converted to the trimethylammonium phosphate **3**. Similarly, diol **42b** with the *myo*-configuration, was converted to phosphate **6**, *via* **46**.

    3. Compounds with Ionic Groups at C(4) and C(5), but without Homo-logation. To complete our series of compounds as potential PLC inhibitors, we finally esterified alcohol **37** with palmitoyl chloride (**37**---->**47**), deprotected the 4,5 positions (**47**---->**48**), and sulfated or phosphorylated the exposed OH groups. In this way, compounds **9** and **10**, analogous to **5** and **6**, but *without* homologation at C(1), were prepared *via* **48**---->**49** and **48**---->**50**, respectively (*Scheme 5*).

    In the present study, a number of analogues of PI and PIP_2 were synthesised and studied for their ability to inhibit PLC. They include inositol derivatives with or without homologation at C(1) and with or without ionic groups (phosphate or sulfate) at C(4) and C(5). In all these compounds, palmitate esters were introduced in place of the

Scheme 3

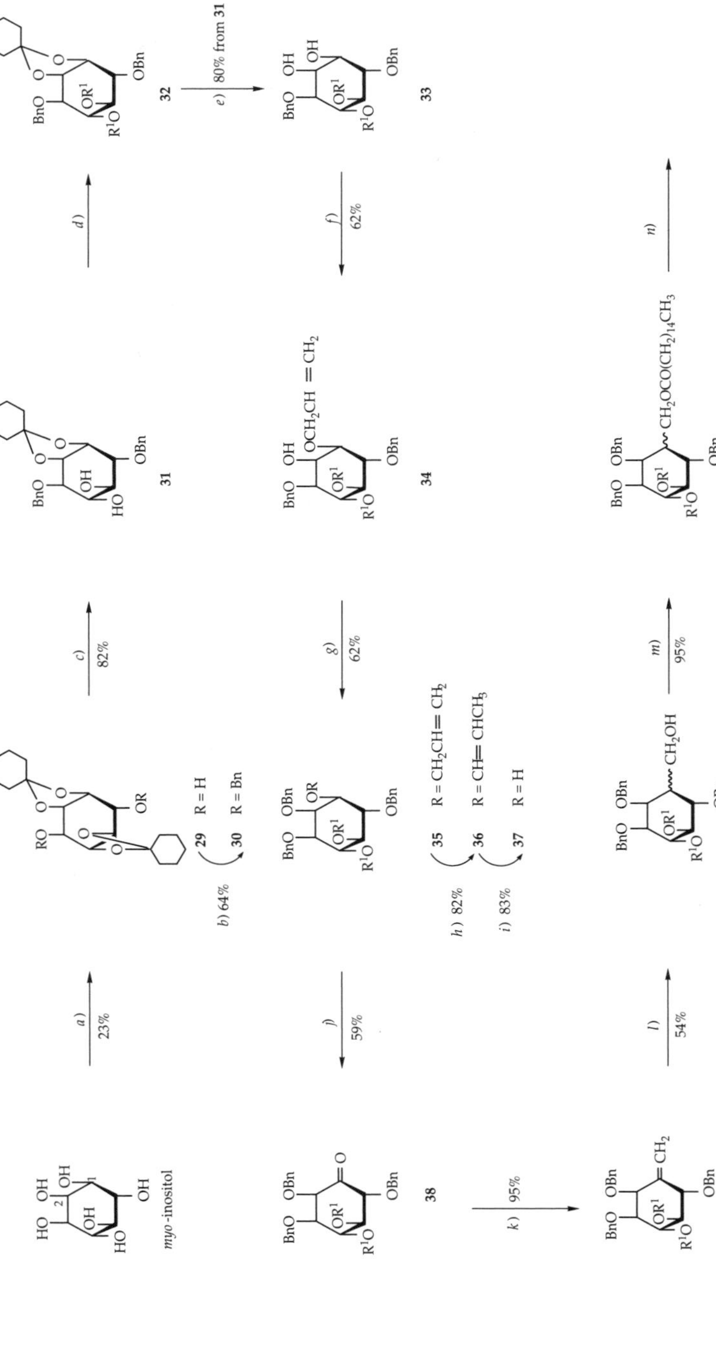

Continued on next page

Scheme 3. Continued.

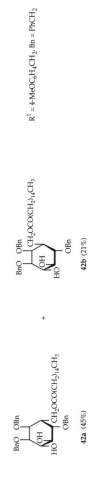

42a (45%)   +   **42b** (21%)

$R^1 = 4\text{-MeOC}_6H_4CH_2$, Bn $= PhCH_2$

a) Cyclohexanone, TsOH, DMF, toluene, 143°, 12 h. *b*) PhCH$_2$Cl, KOH, 100–125°, 3$^1/_2$ h. *c*) Ethane-1,2-diol, CH$_2$Cl$_2$, TsOH, 22°, 30 min. *d*) 4-MeOC$_6$H$_4$CH$_2$Cl, NaH, DMF, 25–60°, 1 h. *e*) Toluene/EtOH/1 M HCl(aq.) 3:6:1, 60°, 2$^1/_2$ h. *f*) 1) Bu$_2$SnO, toluene, 100°, $^3/_4$ h; 2) CH$_2$=CHCH$_2$Br, DMF, 95°, 1$^1/_2$ h. *g*) PhCH$_2$Br, NaH, DMF, 25–40°, 2 h. *h*) PhCH$_2$S, toluene, 25–68°, 1$^1/_2$ h; 2) *t*-BuOK, DMSO, 50°, 19 h. *i*) Toluene/EtOH/1 M HCl(aq.) 3:6:1, 23°, 8$^1/_2$ h. *j*) Pyridinium chlorochromate, CH$_2$Cl$_2$, 25°, 19 h. *k*) CH$_3$(Ph)$_3$PBr, BuLi, THF, 0–22°, 40 min. *l*) 1) BH$_3$·Me$_2$S, toluene, 25–68°, 1$^1/_2$ h; 2) 5 M NaOH, EtOH, 30% H$_2$O$_2$, 10 min, 60°. *m*) CH$_3$(CH$_2$)$_{14}$COCl, pyridine, 4-(dimethylamino)pyridine, CH$_2$Cl$_2$, 25°, 1 h. *n*) DDQ, CH$_2$Cl$_2$, H$_2$O, 25°, 1$^1/_2$ h.

Scheme 4

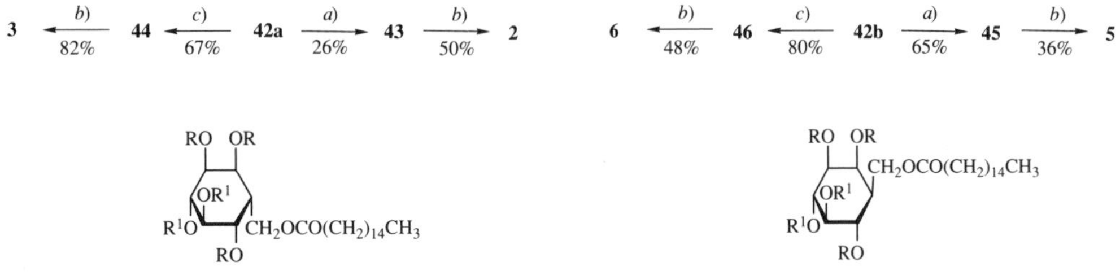

$$3 \xrightarrow[82\%]{b)} 44 \xrightarrow[67\%]{c)} 42a \xrightarrow[26\%]{a)} 43 \xrightarrow[50\%]{b)} 2$$

$$6 \xrightarrow[48\%]{b)} 46 \xrightarrow[80\%]{c)} 42b \xrightarrow[65\%]{a)} 45 \xrightarrow[36\%]{b)} 5$$

42a   R = Bn, R^1 = H
43   R = Bn, R^1 = SO$_3$Na
2   R = H, R^1 = SO$_3$Na
44   R = Bn, R^1 = P(O)(BnO)$_2$
3   R = H, R^1 = P(O)(OH)$_2$·1.6Me$_3$N

42b   R = Bn, R^1 = H
45   R = Bn, R^1 = SO$_3$Na
5   R = H, R^1 = SO$_3$Na
46   R = Bn, R^1 = P(O)(BnO)$_2$
6   R = H, R^1 = P(O)(OH)$_2$·1.6Me$_3$N

a) SO$_3$/pyridine complex, DMF, 25°, 18-20 h. *b)* Pd/C, 1 atm H$_2$, EtOH, 25°, 2-6 h. *c)* 1) 1-*H*-Tetrazole, P(BnO)$_2$[(i-Pr)$_2$N], CH$_2$Cl$_2$, 25°, 2-2$^1/_2$h; 2) 3-ClC$_6$H$_4$CO$_3$H, CH$_2$Cl$_2$, -40° to +25°, 1-1$^1/_2$h.

Scheme 5

$$37 \xrightarrow[91\%]{a)} 47 \xrightarrow[82\%]{b)} 48$$

$$48 \xrightarrow[47\%]{c)} 49 \xrightarrow[8\%]{e)} 9$$

$$48 \xrightarrow[74\%]{d)} 50 \xrightarrow[64\%]{f)} 10$$

Bn = PhCH₂

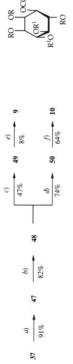

47 R = Bn, R¹ = 4-MeOC₆H₄CH₂

48 R = Bn, R¹ = H

49 R = Bn, R¹ = SO₃(½Ba)

9 R = H, R¹ = SO₃Na

50 R = Bn, R¹ = P(O)(BnO)₂

10 R = H, R¹ = P(O)(OH)₂ · 2.5Me₃N

a) CH₃(CH₂)₁₄COCl, pyridine, 4-(dimethylamino)pyridine, CH₂Cl₂, 25°, 21 h. *b)* DDQ, CH₂Cl₂, H₂O, 25°, 1 h. *c)* SO₃/pyridine complex, DMF, 25°, 14 h, BaCO₃. *d)* 1) 1*H*-Tetrazole, P(BnO)₂[(i-Pr)₂N], CH₂Cl₂, 25°, 2 h; 2) 3-ClC₆H₄CO₃H, CH₂Cl₂, ca. 35°, 30 min. *e)* 1) Pd/C, 1 atm H₂, EtOH, 25°, 2 h; 2) *Dowex 50 W* (H⁺); 3) NaOH. *f)* 1) Pd/C, 1 atm H₂, EtOH, 25°, 2 h; 2) Me₃N.

diacylglyceryl group of PI or PIP$_2$. Only the compounds with ionic groups showed significant activity *in vitro*. However, none of the compounds were able to inhibit PLC in intact cells. As the lack of *in vivo* activity might be due to poor cell penetration, present work is concentrating on the synthesis of prodrugs in order to improve the cellular uptake.

LITERATURE CITED

[1]   Hokin, L.E. *Ann. Rev. Biochem.* **1985**, *54,* 205.

[2]   Berridge, M.J. *Biochem. J.* **1984**, *220,* 345.

[3]   Osborne, N.N.; Tobin, A.B.; Ghazi, H. *Neurochem. Res.* **1988**, *13,* 177.

[4]   Downes, C.P. *Biochem. Soc. Trans.* **1989**, *17,* 259.

[5]   Samuelson, B.; Goldyne, M.; Granström, E.; Hamberg, M.; Hammarström, S.; Malmsten, C. *Ann. Rev. Biochem.* **1978**, *47,* 997.

[6]   Berridge, M.J. *Biotechnology* **1984**, 541.

[7]   International Union of Pure and Applied Chemistry "Nomenclature of Cyclitols", *Pure Appl. Chem.* **1974**, *37,* 283.

[8]   Agranoff, B.W.; Eisenberg Jr., F.; Hauser, G.; Hawthorne, J.N.; Michell, R.H. in "*Inositol and Phosphoinositides*"; Bleasdale, J.E.; Eichberg, J.; Hauser, G., Eds.; Humana Press: Clifton, New Jersey, 1984, p. xxi.

[9]   Hosang, M.; Rouge, M.; Wipf, B.; Eggimann, B.; Kaufmann, F.; Hunziker, W. *J. Cell. Physiol.* **1989**, *140,* 558.

[10] Massy, D.J.R.; Wyss, P. *Helv. Chim. Acta* **1990**, *73,* 1037.

[11] Nashed, M.A.; Anderson, L. *Tetrahedron Lett.* **1976**, 3503.

[12] Tanaka, T.; Tamatsukuri, S.; Ikehara, M. *Tetrahedron Lett.* **1986**, 199.

[13] Perich, J.W.; Johns, R.B. *Synthesis* **1988**, 142.

[14] Yu, K.-L.; Fraser-Reid, B. *Tetrahedron Lett.* **1988**, 979.

RECEIVED February 11, 1991

Chapter 16

Synthesis and Biological Evaluation of Inositol Phospholipid Analogues

John G. Ward and Rodney C. Young

SmithKline Beecham Pharmaceuticals, The Frythe, Welwyn,
Hertfordshire AL6 9AR, England

Close structural analogues of phosphatidylinositol have been prepared by newly-developed synthetic procedures. These compounds have been used to study the requirements for binding to the active sites of a phosphatidylinositol 4-kinase and a phosphoinositide-specific phospholipase C in the search for potential inhibitors. For both enzymes, inhibitory activity is found in compounds possessing at least one lipophilic chain, while the inositol moiety is not essential. Inhibition only of the kinase appears to require a charged linking group.

The importance of the inositol-containing phospholipids in transmembrane cell signalling is now well-established. The least abundant of these, phosphatidylinositol 4,5-bisphosphate [PtdIns(4,5)P$_2$], is cleaved by a receptor-coupled phosphoinositide-specific phospholipase C upon stimulation by a wide variety of neurotransmitters, hormones and growth factors, to release the intracellular second messengers, inositol 1,4,5-trisphosphate [Ins(1,4,5)P$_3$] and sn-1,2-diacylglycerol (DG) *(1)* (Figure 1). Replenishment of the pool of PtdIns(4,5)P$_2$ occurs by the sequential phosphorylation of PtdIns at the 4- and 5- positions, and is catalyzed first by a PtdIns 4-kinase and then by a PtdIns4P 5-kinase. The three classes of phospholipid-handling enzymes mentioned above may be regarded as valid targets in the search for potential therapeutic agents which may be useful in controlling a number of physiological processes, including inflammation, smooth muscle contractility, secretion and cell proliferation *(2,3)*.

A logical starting point in designing inhibitors of any of these enzymes would be to compare the effects of close structural analogues of the natural phospholipid substrates. This would require synthetic methods for the preparation of pure, well-characterized compounds which, until recently, have been unavailable. Although various methods have been reported for the synthesis of PtdIns analogues since the pioneering work of Shvets et al in 1970 *(4)*, many of the stages, including resolution of the inositol building blocks in the case of the chiral phospholipids, were hampered by low yields and lack of reproducibility.

0097–6156/91/0463–0214$06.00/0
© 1991 American Chemical Society

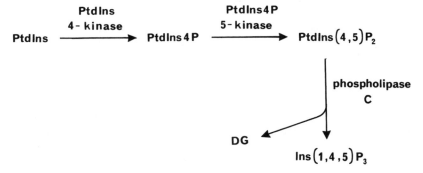

Figure 1. The three classes of phospholipid-handling enzymes.

Chemistry

In all syntheses of inositol phospholipids, three main stages may be identified. First, the preparation of a suitably protected inositol precursor, which must also be resolved into its enantiomers if chirally pure lipids are desired. Then follows a coupling/phosphorylation step (or steps in the cases of PtdIns4P and PtdIns(4,5)P$_2$ syntheses). Finally, deprotection is effected to unmask the required product.

Much of the early work in this field has been reviewed by Gigg *(5)* and here it will be sufficient to summarise very briefly the various approaches to these three stages which have been reported, and described fully in the above mentioned review.

Typical of the inositol protecting groups which have been employed are $\underline{O}$-acetyl, $\underline{O}$-cyclohexylidene and $\underline{O}$-benzoyl. The phosphate group has been protected most frequently as a phenyl, benzyl or 2,2,2-trichloroethyl ester. One may, however, envisage problems of acyl migration in the use of carboxylic esters as protecting groups for the inositol hydroxyls.

Protected inositols have been resolved by forming glycosides (mannosides and glucosides) of, for example, 1(3), 4(6), 5, 6(4)-tetra-$\underline{O}$-benzyl-$\underline{myo}$-inositol, by reaction with mannose or glucose t-butyl orthoacetates. We have found these reactions to be somewhat problematic, since the yields of the glycosylation can be variable, and also since chromatographic separation of the diastereoisomers formed is not straightforward. More recently, chiral esters such as (-)-menthoxyacetate *(6)* and (-)-camphanate *(7)* have been used, but although the esterification of a racemic inositol with the corresponding acid chloride is trivial, the ensuing chromatographic separations can, in some cases, be far from so.

The simplest coupling of a diacyl glycerol and a pentaprotected inositol may be carried out by condensation of these components with phosphorus oxychloride or phenyl phosphorodichloridate. The sometimes rather low yields and long reaction times limit the general usefulness of this method. This coupling may also be effected by reacting a protected inositol having a free 1-OH, with a phosphatidic acid, in the presence of an activating agent such as triisopropylbenzene sulfonyl chloride (TPS) or mesitylene sulfonyl chloride. The coupling may also be carried out in the "reverse" sense by condensing a protected inositol 1-phosphomonoester with a diacyl glycerol in the presence of the same activating agents. We have found that this approach again gives only moderate yields (~40%) and requires long reaction times. A significant yield of the sulfonate ester of the inositol is also frequently obtained.

Another means of effecting this coupling involves reaction of protected inositol 1-$\underline{O}$-benzylphosphate silver salts with 1,2-di-$\underline{O}$-acyl-$\underline{sn}$-glycerol-3-iodohydrins. This seems to give acceptable yields and the benzylphosphate protecting group renders this route applicable to the synthesis of lipids having unsaturation in the fatty acid chains, since it may be removed by treatment with sodium iodide.

The final deprotection of a protected lipid has been effected in various ways, depending on the chemical nature of the protecting groups used. Benzyl groups may be removed by hydrogenolysis over palladium, although the complete debenzylation of polybenzylated inositols has proved problematic in this laboratory. Phenyl phosphate esters may be efficiently removed by hydrogenolysis in the presence of Adams' catalyst. The cleavage of 2,2,2-trichloroethyl phosphates appears to be somewhat capricious and depends very much on the activity of the zinc dust used. Cyclic ketal (cyclohexylidene or isopropylidene) protecting groups employed to block the inositol hydroxyls are normally left in place until the protected phosphate linkage is cleaved, when an aqueous or aqueous/ethanolic solution of the resulting phosphodiester is sufficiently acidic to effect cleavage.

This brief summary suggests that several of the stages described could usefully be improved. It was already known from work carried out in these laboratories that racemic 2,3(1):5,6(4)-di-O-isopropylidene-myo-inositol could be silylated selectively in the 1(3)-position, and the resulting racemic silyl ether was chosen as a suitable precursor for optical resolution (Scheme 1). Initially the racemate (±)-1 was resolved by formation of the diastereoisomeric (-)-camphanate esters, 2a and 3a (8) but this approach required an extremely difficult and tedious chromatographic separation. Subsequently, it was found that the corresponding diastereoisomers 2b and 3b formed by reaction of (±)-1 with the acid chloride of N-(2,2,2-trichloroethoxycarbonyl)-(R)-proline were far more easily separated (9). Correlation of the optical rotation of the product resulting from alkaline hydrolysis of the less polar diastereoisomer, 2b, with that of compound (-)-1 prepared by alkaline hydrolysis of its optically pure (-)-camphanate ester (the absolute configuration of which had been determined by single crystal X-ray analysis (10)) showed them to be identical. The silyl ethers (-)-1 and (+)-1 could then be converted, into the 4-methoxytetrahydropyranyl (MTHP) derivatives (+)-4 or (-)-4 (11), or alternatively desilylated to give the optically active diacetonides (+)-5 and (-)-5.

It was found possible to couple (+)-4 with the phenylphosphatidic acid (+)-6 (as its Na salt) rapidly and in good yield (10) giving 7, by employing 1-(mesitylene-2-sulfonyl)-3-nitro-1,2,4-triazole (MSNT), a phosphate activating reagent originally developed in the field of ribonucleotide synthesis (12)(Scheme 2). (See Chart 1.)

In a like manner (+)-6 was coupled with (-)-4, and (-)-6 was coupled with (+)-4 and (-)-4 yielding the three fully protected lipids stereoisomeric with 7. It subsequently proved possible to selectively phosphorylate the acetonide (+)-5 at the 1-OH group to give 8, in 60% yield (9), thus obviating the need for protection of the 4-OH. This coupling is a key step in syntheses of other PtdIns analogues such as 1-hexadecylphosphoryl-D-myo-inositol 14, which results (after deprotection) from the MSNT mediated coupling of hexadecylphenylphophate and the acetonide (+)-5. The phenylphosphate ester in 7 or 8 could then be cleaved by hydrogenolysis over PtO$_2$, and the analogous stereoisomeric protected lipids were cleaved in exactly the same fashion. (See Chart 2.)

Aqueous ethanolic solutions of the resulting phosphodiesters were allowed to stand at ~40°C, whereby the inositol -OH groups were deprotected. PtdIns analogues were normally converted into the Na salts (9a-d) for increased stability. (See Chart 3.)

The PtdIns4P analogue 11 was synthesized (9) by phosphitylating the intermediate 8 using diphenyl N,N-diethylphosphoramidite in the presence of 1H-tetrazole (Scheme

(Scheme 1)

Reagents: 　i, RCOCl, pyridine; ii, KOH, EtOH.

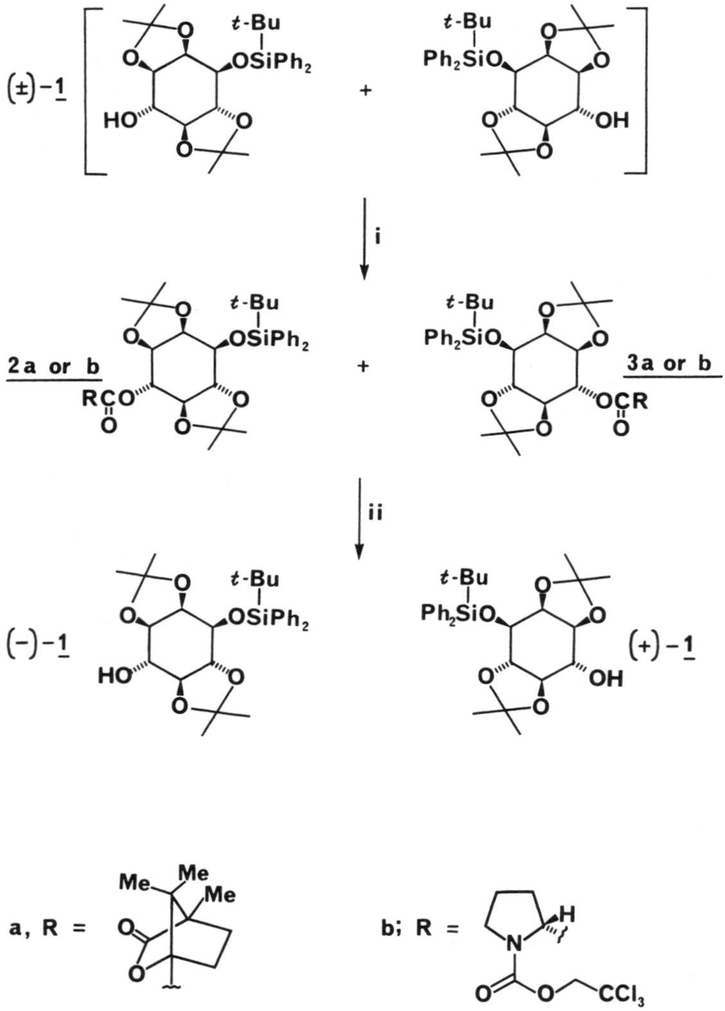

a, R =

b; R =

Chart 1

$(+)-\underline{4}$

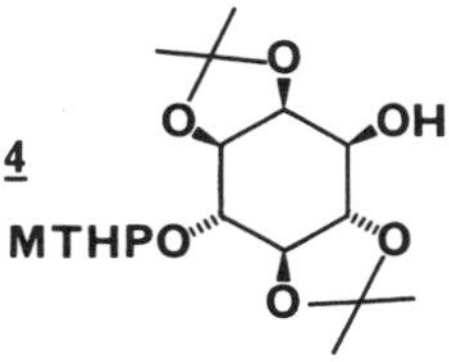

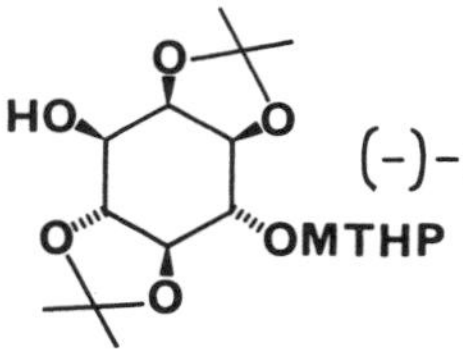

MTHPO

$(-)-\underline{4}$

OMTHP

$(+)-\underline{5}$

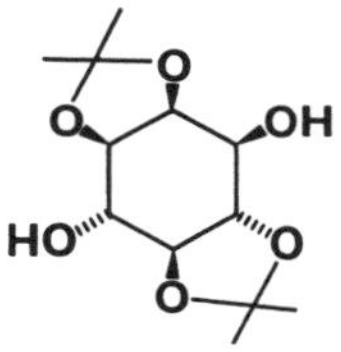

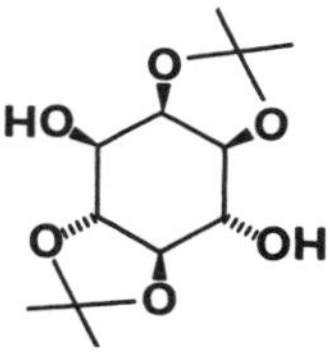

HO

$(-)-\underline{5}$

OH

MTHP =

OMe

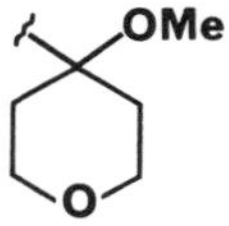

Chart 2

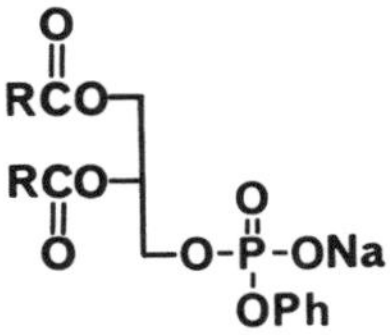

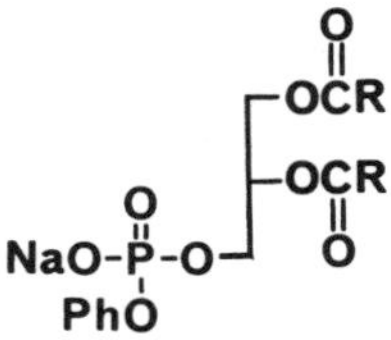

$(+)-\underline{6}$

$(-)-\underline{6}$

$R = -(CH_2)_{14}CH_3$

(Scheme 2)

Reagents: i, MSNT, pyridine;   ii, H$_2$, PtO$_2$, EtOH;
             iii, EtOH, H$_2$O; iv, Amberlite IRC 50 (Na form), EtOH aq.

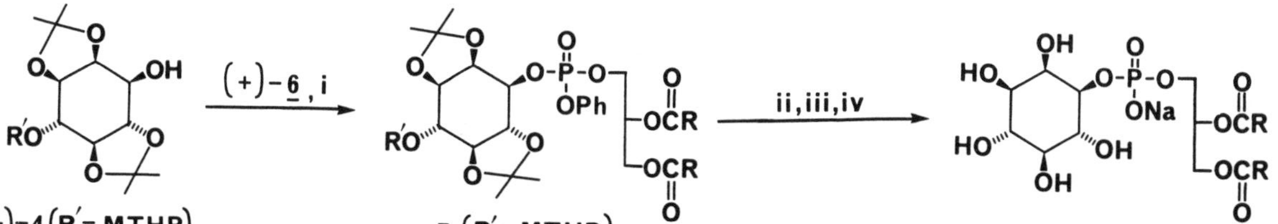

$(+)-\underline{4}(R'=MTHP)$
$(+)-\underline{5}(R'=H)$

$\underline{7}\ (R'=MTHP)$
$\underline{8}(R'=H)$

$\underline{9a}$

R $= -(CH_2)_{14}CH_3$

Chart 3

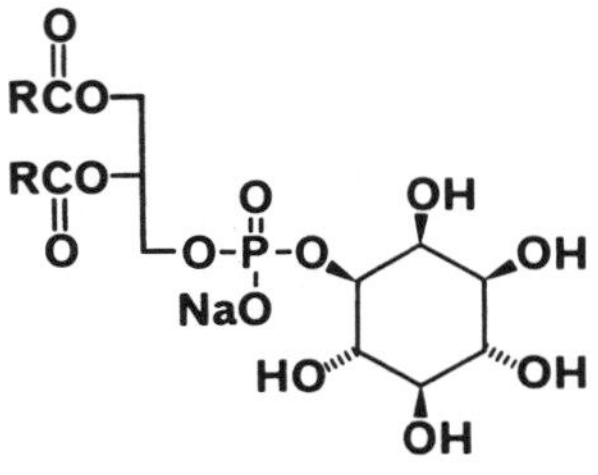

9b

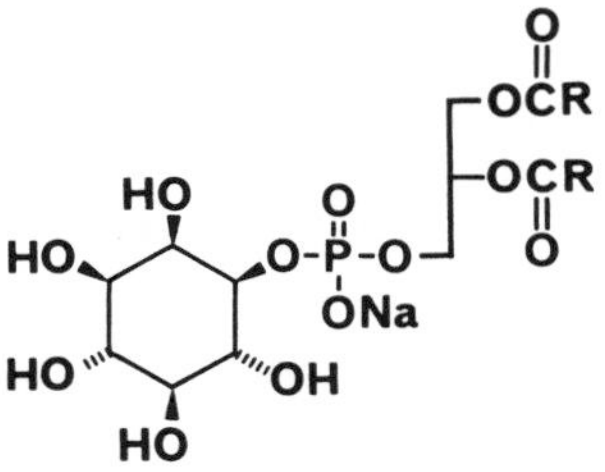

9c

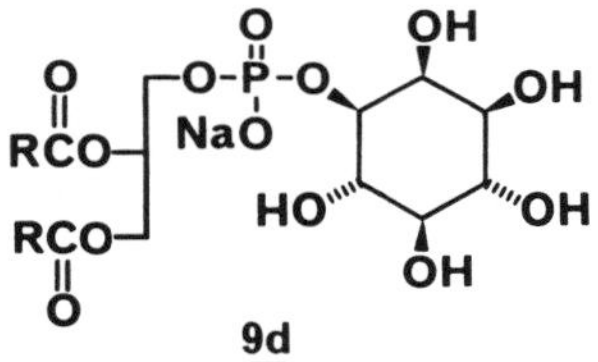

9d

Chart 4

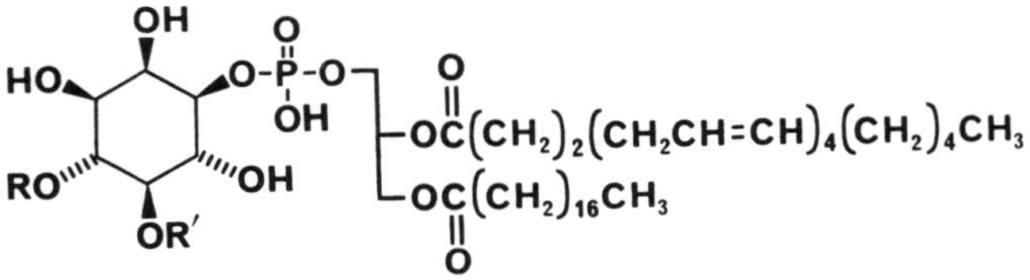

PtdIns	R = R' = H
PtdIns4P	R = PO$_3$H$_2$ R' = H
PtdIns(4,5)P$_2$	R = R' = PO$_3$H$_2$

12	n = 15
13	n = 5

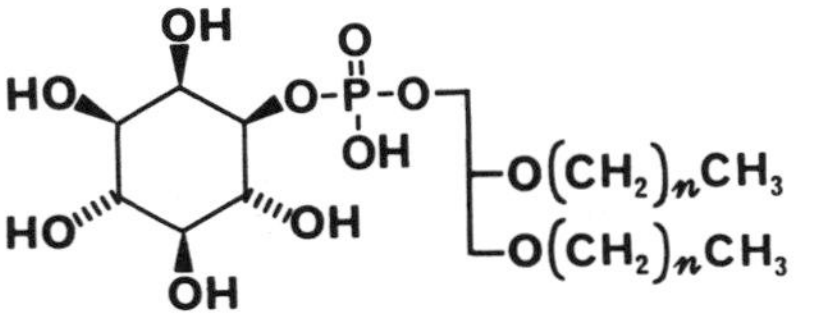

3). The resulting phosphite triester was then oxidized with t-butyl hydroperoxide to give 10 in 73% overall yield. Deprotection was effected by hydrogenolysis over PtO_2 and hydrolysis as detailed above. (See Chart 4.)

The phophoramidite approach has also been used effectively by Van Boom et al *(13)* in a synthesis of a PtdIns(4,5)P$_2$ analogue. In this case the coupling of the diacyl glycerol and inositol portions was also carried out using phosphoramidite chemistry. Using a combination of the methods described above a number of structural variants of PtdIns have also been prepared, the detailed chemistry of which will be reported in a forthcoming publication.

Biological Discussion

The inositol phospholipids contain chiral centres in both the myo-inositol and glycerol moieties, and in order to investigate the stereochemical requirements for inositol phospholipid binding to PtdIns 4-kinase, we first compared the phosphorylation rates of the four possible stereoisomers of synthetic 1,2-dihexadecanoyl PtdIns with that of mammalian PtdIns, using a partially-purified enzyme derived from human erythrocyte membranes. The results, presented in Table 1, show that the mammalian phospholipid, comprising mainly the 1-0-octadecanoyl-2-0-(5,8,11,14-eicosatetraenoyl) PtdIns, phosphorylates at a similar rate to that of its synthetic counterpart, 9a. Moreover, the diastereoisomer of 9a with the inverse stereochemistry in the glycerol moiety (9c) is also an excellent substrate. In contrast, inversion of the stereochemistry in the inositol ring by attachment of the phosphodiester group to the inositol D-3 position, results in compounds 9b and 9d which are poor substrates for this enzyme, phosphorylating at only 3 - 4% of the rate of analogues 9a and 9c having a D-1-substituted inositol *(10)*.

Using mammalian PtdIns as the substrate, the four stereoisomers, 9a - 9d were next compared as inhibitors of PtdIns 4-kinase. The results, presented in Table 1 alongside the substrate activities, show a close parallel with phosphorylation rates, indicating that these compounds are inhibiting the enzyme by specifically binding to the active site, and not by nonspecific disruption of the mixed substrate/detergent micelle in the assay, as might be possible for compounds having such an amphiphilic structure.

These combined results show that the chirality of the inositol ring is crucial for efficient phosphorylation and for effective binding to the enzyme, suggesting that the axial inositol 2-hydroxyl group plays a key role in recognition by the enzyme active site. In addition, the chirality and chemical composition of the glycerol moiety appear to be relatively unimportant, and it is likely that the diacylglycerol moiety serves only as a nonspecific anchor to the membrane thus reducing the number of degrees of freedom for encountering the membrane bound enzyme *(10)*.

With this information in hand, we extended our investigation of the structural requirements for PtdIns 4-kinase inhibition by synthesizing a number of analogues of PtdIns having structural modifications in the inositol, diacylglycerol and phosphate portions of the molecule, and comparing their inhibitory effects on the

(Scheme 3)

Reagents: i, (PhO)$_2$PNEt$_2$, tetrazole, CH$_2$Cl$_2$; ii, Et$_3$N, t-BuOOH;
iii, H$_2$, PtO$_2$, EtOH-EtOAc (83:17); iv, EtOH, EtOAc,
H$_2$O (70:15:15), 25 → 40°C; v, 25% NH$_3$ aq.

10

11 $\left(NH_4^+ \text{ salt}\right)$

Table 1 **Comparison of PtdIns Analogues as Substrates and Inhibitors for Human Erythrocyte PtdIns 4-kinase**

Compound	Substrate Activity[a]	Inhibitor Activity[b]
PtdIns	100	55.3±1.4% (IC$_{50}$=363±49 μM)
9a	106	46.2±1.5%
9b	4.3	8.6±2.6%
9c	100	31.5±2.7%
9d	3.2	8.0±3.2%

a. Phosphorylation rate relative to PtdIns (= 100)
b. Percentage inhibition at 500 μM

catalysis of mammalian PtdIns phosphorylation.Alterations to the diacylglycerol moiety in PtdIns 9a revealed a high degree of tolerance in the enzyme. The long acyl chains in 9a could be replaced by ether chains of comparable length to give a compound 12, which retains inhibitory activity, although the racemic analogue with the shorter hexyl chains 13 is significantly less active. Furthermore, the glycerol moiety is not an essential structural component. Simple D-1-inositol alkyl phoshates were found to inhibit the kinase reasonably well, provided that the hydcarbon chain was at least 15 carbon atoms long, for example, the hexadecyl analogue, 14 *(14)* (Table 2). These results are important for two reasons. First, they show the relative insensitivity to structural modifications in the diacylglycerol portion of PtdIns and support the suggestion that this molecular feature is important only in functioning as a membrane anchor. Second, the potentially unstable diacylglycerol moiety is not essential for binding, and can be replaced by glycerol ethers or single alkyl chains. Thus, problems of synthesis, handling and storage of analogues can be correspondingly simplified.

Attempts to retain inhibitory activity while replacing the phosphate diester group in (14) were unsuccessful. For example, the uncharged hexadecyl methane phosphonate, (15) and hexadecylsulfonate, (16), have no effect on the enzyme in concentrations of up to 500 μM. These results suggest that a coulombic interaction is important in binding a ligand to the active site, and the apparent need for a charged phosphate moiety may lead to serious difficulties in ultimately designing a cell-penetrant inhibitor.

Investigation of the scope for substitution and replacement of the myo-inositol ring in analogues of 9a and 14 was more rewarding. Substitution at the ring 4- and 5-positions of mammalian PtdIns to give PtdIns4P and PtdIns $(4,5)P_2$ had little effect on enzyme inhibition. Thus, for PtdIns 4-kinase, enzyme activity may be regulated by product inhibition and in concentrations close to the K_m for PtdIns (Km = 116 μM). Similarly, little effect on inhibitory potency was observed on introducing a phosphate group into the inositol 4-positions of 9a and 14 (in 17 and 18, respectively), again supporting a common mode of binding to the enzyme. Interestingly, introduction of a neutral diethylphosphate group at the 4-position of 14 (in 19) apparently leads to a small reduction in binding to the enzyme. (See Chart 5.) A myo-inositol ring system is not essential in an inhibitor of PtdIns 4-kinase. Retention of only the inositol 4-hydroxyl group in 9a or 14, or the axial 2-hydroxyl group in 14 gave compounds (20, 21 and 22 respectively) of comparable potency to the parent compounds. This result infers that none of the hydroxyl groups are actually involved in binding to the enzyme active site. The poor binding of the PtdIns stereoisomers with 'unnatural' stereochemistry in the inositol moiety (9b and 9d) must, therefore, be rationalized in terms of an undesirable interaction with the enzyme, perhaps steric repulsion involving the axial 2-hydroxyl group. (See Chart 6.) Further progress towards a potent, cell-penetrant inhibitor of PtdIns 4-kinase may now be made by incorporating these structural features into a bisubstrate analogue inhibitor using our knowledge of the requirements for inhibitor binding to the ATP site *(15)*.

Having synthesized a number of close structural analogues of PtdIns, we next

Table 2 Inhibitory Activities of PtdIns Analogues on Human Erythrocyte PtdIns 4-kinase and Guinea Pig Uterus PtdIns-phospholipase C

Compound	PtdIns 4-kinase[a]	Phospholipase C[b]
PtdIns	55.3±1.4%(IC_{50}=363±49μM)	40 μM
PtdIns 4P	58.8±2.7%(IC_{50}=469± 9μM)	
PtdIns(4,5)P_2	75.4±1.8%(IC_{50}=267± 4μM)	
12	46.8±0.9	
13[c]	14.7±1.8	>300 μM
14	34.7±4.6%(K_i=360±12μM)[d]	54 μM
15[c]	0%	
16[c]	0%	120 μM
17	42.0±4.6%	
18[c]	54.8±2.1%(IC_{50}=253±101μM)	98 μM
19[c]	23.6±2.2%	
20	47.3±1.8%	
21	31.3±1.8%	25 μM
22	36.6±3.4%(K_i=570±20μM)[d]	>300 μM
23[c]	0%	28 μM
24[c]		31 μM

a. Percentage inhibition at 500 μM. b. IC_{50} from single experiments.
c. Racemic in Ins moiety. d. Competitive inhibition established.

Chart 5

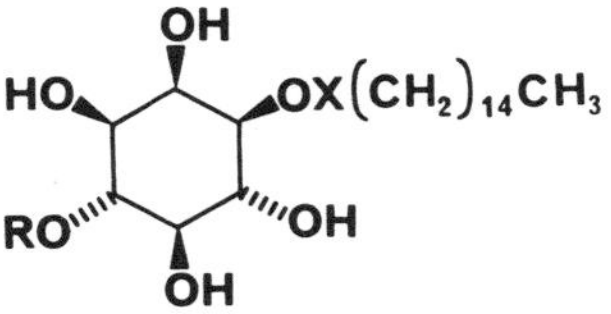

14	X = P(=O)(OH)-OCH$_2$	R = H
15	X = P(=O)(CH$_3$)-OCH$_2$	R = H
16	X = SO$_2$CH$_2$	R = H
18	X = P(=O)(OH)-OCH$_2$	R = P(=O)(OH)$_2$
19	X = P(=O)(OH)-OCH$_2$	R = P(=O)(OEt)$_2$
23	X = CO	R = H
24	X = CH$_2$	R = H

Chart 6

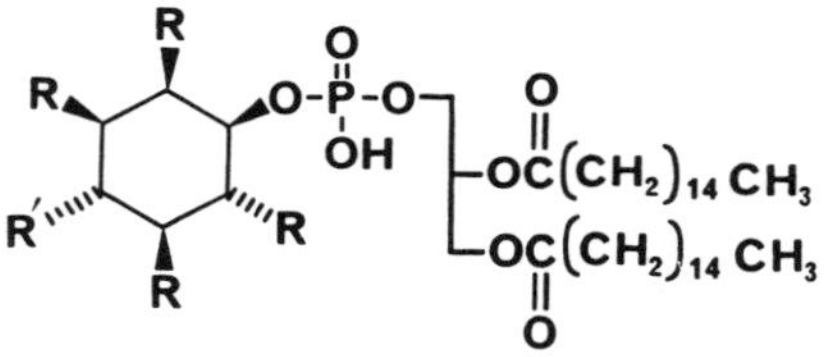

| 17 | R = OH | R′ = OP(OH)₂ |
| 20 | R = H | R′ = OH |

| 21 | R = H | R′ = OH |
| 22 | R = OH | R′ = H |

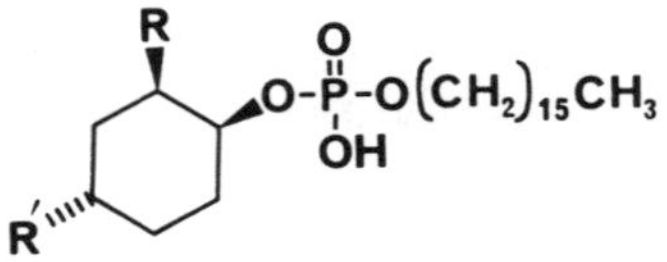

considered the possiblity that these compounds might interfere with the production of second messengers by inhibiting the hydrolysis of PtdIns $(4,5)P_2$. The effects of selected compounds were examined using a partially-purified phosphoinositide-specific phospholipase C obtained from guinea pig uterus. Thus far, comparisons have only been made between IC_{50} values from single experiments using mammalian PtdIns as substrate *(16)*. As was found with inhibition of PtdIns 4-kinase, an essential structural requirement for an inhibitor appears to be a lipophilic tail, composed of one or two hydrocarbon chains. Again, chain length is important, and the single hexadecyl group in 14 appears to be equieffective to a diacylglyceryl moiety, as, for example, in mammalian PtdIns. Conversely, however, a charged phosphate group is not required for effective inhibition. Thus, the carboxylic ester (23) and the ether (24) are at least as potent as the phosphate ester (14), having IC_{50} values of 28, 31 and 54 μM respectively. Assuming competitive inhibition, K_i values for 24 and 14 can be calculated as 12 and 14 μM, respectively, which are close to the K_m of PtdIns (25 μM) for this enzyme.

The inositol ring is apparently not essential for recognition and binding. Interestingly, retention of only the inositol 4-hydroxyl group (in 21) results in a compound of approximately comparable potency to 14, while retention of only the axial 2-hydroxyl group (in 22) leads to a dramatic loss of potency, suggesting that the 2,3,5 and 6-hydroxyl groups are not involved in binding.

In summary, human erythrocyte membrane PtdIns 4-kinase and guinea pig uterus PtdIns-phospholipase C show both similarities and differences in their sensitivity to inhibition by PtdIns analogues. For both enzymes, a structural requirement for inhibition is at least one lipophilic chain of about 15 carbon atoms, while possession of a myo-inositol moiety is not essential. The kinase was inhibited only by compounds which incorporate a charged linking group into their structure, whereas this feature does not appear to be important for inhibition of the lipase.

Literature Cited

1.   Downes, C.P.; Michell, R.H. In *Molecular Aspects of Cellular Regulation,* vol.4 *Molecular Mechanisms of Transmembrane Signalling;* Cohen, P.; Houslay, M.D., Eds.; Elsevier:Amsterdam, **1985**; p.3.

2.   Berridge, M.J.; Irvine, R.F. *Nature (London)* **1984**, 312, 315.

3.   Nishizuka, Y. *Nature (London)* **1984**, 308.

4.   Zhelvakova, E.G.; Klyashchitskii, B.A.; Shvets, V.I.; Evstigneeva, R.P.; Preobrazhenskii, N.A. *Zh. Obshch. Khim.* **1970**, 40(1), 248.

5.   Gigg, R. *Chem. Phys. Lipids* **1980**, 26, 287.

6.   Ozaki, S.; Watanabe, Y.; Ogasawara, T.; Kondo, Y.; Shiotani N; Nishii, H.; Matsuki, T. *Tetrahedron Lett.* **1986**, 27, 3157.

7.   Billington, D.C.; Baker, R.; Kulagowski, J.J.; Mawer, I.M. *J. Chem. Soc. Chem. Commun.* **1987**, 314.

8.   Ward, J.G.; Young, R.C. *Tetrahedron Lett.* **1988**, 29, 6013.

9.   Jones, M.; Rana, K.K.; Ward, J.G.; Young, R.C. *Tetrahedron Lett.* **1989**, <u>30</u> 5353.

10. Young, R.C.; Downes, C.P.; Eggleston, D.S.; Jones, M.; Macphee, C.H.; Rana, K.K.; Ward, J.G. *J. Med. Chem.* **1990**, <u>33</u>, 641.

11. Reese, C.B.; Saffhill, R.; Sulston, J.E. *Tetrahedron.* **1970**, <u>26</u>, 1023.

12. Jones, S.S.; Rayner, B.; Reese, C.B.; Ubasawa, A.; Ubasawa, M. *Tetrahedron.* **1980**, <u>36</u>, 3075.

13. Dreef, C.E.; Elie, C.J.J.; Hoogerhout, P.; Van der Marel, G.A.; Van Boom, J.H. *Tetrahedron Lett.* **1988**, <u>29</u>, 6513.

14. Young, R.C.; Downes, C.P. *Drug Design and Delivery.* **1990**, <u>6</u>, 1.

15. Young, R.C.; Jones, M.; Milliner, K.J.; Rana, K.K.; Ward, J.G. *J. Med. Chem.* **1990**, <u>33</u>, 2073.

16. Bennett, C.F.; Angioli, M. Unpublished studies.

RECEIVED March 11, 1991

Author Index

Affiliation Index

Subject Index

Production: Margaret J. Brown
Indexing: Deborah H. Steiner
Acquisition: A. Maureen Rouhi
Cover design: Thomas Ford—Holevinski and Sue Schafer

Printed and bound by Maple Press, York, PA

Other ACS Books